Nutrient Requirements of Horses

Fifth Revised Edition, 1989

Subcommittee on Horse Nutrition
Committee on Animal Nutrition
Board on Agriculture
National Research Council

NATIONAL ACADEMY PRESS
Washington, D.C.   1989

National Academy Press   2101 Constitution Avenue, NW   Washington, DC 20418

NOTICE: The project that is the subject of this report was approved by the Governing Board of the National Research Council, whose members are drawn from the councils of the National Academy of Sciences, the National Academy of Engineering, and the Institute of Medicine. The members of the committee responsible for the report were chosen for their special competences and with regard for appropriate balance.

This report has been reviewed by a group other than the authors according to procedures approved by a Report Review Committee consisting of members of the National Academy of Sciences, the National Academy of Engineering, and the Institute of Medicine.

The National Academy of Sciences is a private, nonprofit, self-perpetuating society of distinguished scholars engaged in scientific and engineering research, dedicated to the furtherance of science and technology and to their use for the general welfare. Upon the authority of the charter granted to it by the Congress in 1863, the Academy has a mandate that requires it to advise the federal government on scientific and technical matters. Dr. Frank Press is president of the National Academy of Sciences.

The National Academy of Engineering was established in 1964, under the charter of the National Academy of Sciences, as a parallel organization of outstanding engineers. It is autonomous in its administration and in the selection of its members, sharing with the National Academy of Sciences the responsibility for advising the federal government. The National Academy of Engineering also sponsors engineering programs aimed at meeting national needs, encourages education and research, and recognizes the superior achievements of engineers. Dr. Robert M. White is president of the National Academy of Engineering.

The Institute of Medicine was established in 1970 by the National Academy of Sciences to secure the services of eminent members of appropriate professions in the examination of policy matters pertaining to the health of the public. The Institute acts under the responsibility given to the National Academy of Sciences by its congressional charter to be an adviser to the federal government and, upon its own initiative, to identify issues of medical care, research, and education. Dr. Samuel O. Thier is president of the Institute of Medicine.

The National Research Council was organized by the National Academy of Sciences in 1916 to associate the broad community of science and technology with the Academy's purposes of furthering knowledge and advising the federal government. Functioning in accordance with general policies determined by the Academy, the Council has become the principal operating agency of both the National Academy of Sciences and the National Academy of Engineering in providing services to the government, the public, and the scientific and engineering communities. The Council is administered jointly by both Academies and the Institute of Medicine. Dr. Frank Press and Dr. Robert M. White are chairman and vice chairman, respectively, of the National Research Council.

This study was supported by the Agricultural Research Service of the U.S. Department of Agriculture, under Agreement No. 59-32U4-5-6, and by the Center for Veterinary Medicine, Food and Drug Administration of the U.S. Department of Health and Human Services, under Cooperative Agreement No. FD-U-000006-06-1. Additional support was provided by the American Feed Industry Association, Inc. Any opinions, findings, conclusions, or recommendations expressed in this publication are those of the authoring subcommittee and do not necessarily reflect the views of the sponsors.

Library of Congress Cataloging-in-Publication Data
National Research Council (U.S.). Subcommittee on Horse Nutrition.
    Nutrient requirements of horses/Subcommittee on Horse Nutrition,
    Committee on Animal Nutrition, Board on Agriculture, National
    Research Council.—5th rev. ed.
        p. cm.
    Bibliography: p.
    Includes index.
        1. Horses—Nutrition—Requirements.   2. Horses—Feeding and feeds.
    I. Title.
    SF285.5.N37   1989
    636.1′08′52—dc19   89-3061
    ISBN 0-309-03989-4   CIP

Preface

Research on the nutrition and feeding of horses has increased substantially during the 10 years since the last edition. In some cases, this work has served to support and strengthen previous recommendations. In other cases, it has suggested that earlier recommendations were inadequate or excessive. For some nutrients, we now have information where previously none existed. When information was limited or nonexistent, the subcommittee chose to use information available on other species to estimate the horse's needs. Our knowledge of horse nutrition still contains many gaps that the subcommittee hopes will be filled in the near future. The requirements given represent the best information available at this time.

Changes in this edition reflect new information available on the requirements of horses, which are discussed in Chapter 2. For example, in calculating the digestible energy requirements for maintenance, the equation has been changed from a metabolic weight basis to a body weight basis. Also, the equations for calculating digestible energy requirements for work, reproduction, and growth have been revised to reflect new information. Because digestible energy intake has a direct effect on production and performance, where practical the requirements for other nutrients are expressed as a function of digestible energy intake. In addition, nutrient requirements for two rates of growth are calculated for weanlings and yearlings to illustrate the effect of growth rate on nutrient needs.

Chapter 3 discusses characteristics of feedstuffs often used in diets for horses, as well as some health problems that may result with certain types of feedstuffs. General considerations for selecting a diet for foals and for adult and growing horses are discussed in Chapter 4. Chapter 5 presents the mathematical statements necessary to calculate daily requirements for digestible energy, crude protein, lysine, calcium, phosphorus, magnesium, potassium, and vitamin A. These equations were used to develop the computer diskette available with the report.

This edition should be a useful tool and guide for the nutritional management of horses and ponies during various phases of the life cycle.

The Subcommittee on Horse Nutrition was appointed in 1985 under the auspices of the Board on Agriculture's Committee on Animal Nutrition to update and revise the fourth edition of *Nutrient Requirements of Horses*, issued in 1978. This new edition was reviewed by the Committee on Animal Nutrition, the Board on Agriculture, and four outside reviewers. The subcommittee is grateful to these individuals for their efforts. The subcommittee also acknowledges the assistance of Robert van Saun in completing the computer program included with this edition and would like to thank Board on Agriculture staff, Sharon Giduck, staff officer, and Kamar Patel, senior secretary, for their assistance in the preparation of this edition.

SUBCOMMITTEE ON HORSE NUTRITION

EDGAR A. OTT, *Chairman*, University of Florida, Gainesville

JOHN P. BAKER, University of Kentucky

HAROLD F. HINTZ, Cornell University

GARY D. POTTER, Texas A&M University

HOWARD D. STOWE, Michigan State University

DUANE E. ULLREY, Michigan State University

iv

Contents

Tables

Nutrient Requirements of Horses

Fifth Revised Edition, 1989

1 Introduction

The horse industry has become a very important part of the agricultural scene in many areas of the United States. A major increase in research on physiological factors affecting production and performance has resulted from the popularity of the horse and the economic impact of the industry; however, many of the problems have not been solved. The subcommittee found large gaps in the published information, unresolved conflicting reports, and a disconcerting need to apply information gathered under one set of circumstances to very broad and diverse management systems.

The requirements stated herein indicate the minimum amounts needed to sustain normal health, production, and performance of horses. The numbers are not averages of all the data available. They are selected values based on published research, calculations designed to extrapolate information over the total population, and the subcommittee's experience in applying information to field situations. For example, if the only research available on the weanling foal's needs for a specific nutrient was conducted with Thoroughbred weanlings, that information was also used to make recommendations for pony weanlings and draft horse weanlings.

Horses should be fed as individuals. In applying these recommendations, consideration should be given to the following factors:

- digestive and metabolic differences among horses that result in some horses being "hard keepers" and others "easy keepers," and appropriate adjustments in feed intake to compensate for this variation;
- variation in production and performance capabilities of the animal and expectations of the owner;
- health status of the animal;
- variations in the nutrient availability in feed ingredients;
- interrelationships among nutrients;
- previous nutritional status of the horse; and
- climatic and environmental conditions.

1

2 Nutrient Requirements, Deficiencies, and Excesses

ENERGY

Source

CARBOHYDRATES

The horse is a nonruminant herbivore that uses dietary carbohydrate as the primary energy source. In contrast to the situation in ruminants, carbohydrates are exposed to pancreatic and intestinal enzymes before reaching major regions of fermentation (Roberts, 1975a). Nonstructural carbohydrates, such as starch, maltose, and sucrose, are hydrolyzed and absorbed as monosaccharides proximal to the cecum. Lactose can be digested by young horses, but horses older than 3 years appear to have limited lactase activity (Roberts, 1975b). As a consequence, sudden introduction of lactose-containing feedstuffs to mature horses may induce digestive disturbances.

The proportion of nonstructural dietary carbohydrates digested in the small intestine can be very significant for concentrate feedstuffs. For example, Hintz et al. (1971b) reported prececal apparent digestibility values of 71 percent for nonstructural carbohydrates in a high-corn diet and 46 percent for those in a high-alfalfa diet. Glucose derived from this digestion enters the portal vein and contributes in a major way to the horse's energy budget. Nonstructural carbohydrates that escape prececal digestion and structural carbohydrates, such as cellulose and hemicellulose, are subjected to anaerobic microbial fermentation, largely in the cecum and colon. This fermentation produces the volatile fatty acids—acetic, propionic, isobutyric, butyric, isovaleric, and valeric (Hintz et al., 1971a). These volatile fatty acids are absorbed and constitute important sources of energy, particularly for horses fed high-forage diets.

As a consequence of this digestive and metabolic strategy, the horse has plasma glucose concentrations intermediate between those of ruminants and simple stomached omnivores (Evans, 1971). Endocrine regulation of glucose and volatile fatty acid metabolism seems to be quite unlike that in ruminants and more like that of nonherbivores with respect to insulin secretion and sensitivity (Argenzio and Hintz, 1971). However, intravenous infusion of glucose results in removal of exogenous glucose from the blood of ponies at a slower rate than in omnivores, such as man, but faster than in adult ruminants, such as sheep (Mehring and Tyznik, 1970). In addition, the equine pancreas responds slowly to a decline in plasma glucose compared to the pancreas of humans (Madigan and Evans, 1973).

Considerable variation in plasma insulin concentrations among ponies has been noted (Argenzio and Hintz, 1971; Ralston et al., 1979). The cause of this variation has not been determined, but it is not reflected either in the plasma glucose concentration or in feeding responses of ponies. Intravenous glucose infusions delay the initiation of eating by ponies but do not clearly control meal size or duration of eating (Ralston and Baile, 1982). Changes in cecal volatile fatty acid concentration can generate cues that may contribute to control of meal size and frequency (Ralston et al., 1983). Ponies with chronic laminitis appear significantly less sensitive to insulin than do unaffected controls (Coffman and Colles, 1983). This tendency toward impaired tissue sensitivity to insulin may also predispose ponies to hyperlipidemia (Jeffcott and Field, 1985). In addition, obesity may be associated with hyperinsulinemia, hyperglycemia, and laminitis (Corke, 1986; Jeffcott et al., 1986).

The pattern of glucose metabolism is influenced by the character of the diet (Argenzio and Hintz, 1972; Garcia and Beech, 1986). The low percentage of carbon dioxide derived from glucose in ponies fed a high-forage diet closely approximates that expected in ruminants

2

and suggests that a large fraction of metabolic energy comes from volatile fatty acids. The extensive conversion of glucose to carbon dioxide in grain-fed ponies, and the high percentage of carbon dioxide derived from glucose, suggest that such a diet promotes a higher percentage contribution of energy from the oxidative metabolism of glucose, comparable to that in nonherbivorous animals.

Gluconeogenesis from cecal propionate has been demonstrated in ponies fed hay or hay plus wheat bran (Ford and Simmons, 1985). Approximately 7 percent of total glucose production was derived from propionate produced in the cecum, and this percentage was unaffected by diet.

Energy metabolism in the exercising horse, and the role of dietary and tissue carbohydrates in that metabolism, have received attention recently. Muscle glycogen stores can be increased by training (Goodman et al., 1973; Lindholm and Piehl, 1974; Topliff et al., 1983) and by adjusting the diet and training regimens (Topliff et al., 1985). Kline and Albert (1981) fed low-, conventional-, or high-carbohydrate diets to Standardbred horses and produced muscle glycogen concentrations of 107, 130, and 144 mmol/kg of wet tissue weight, respectively. Topliff et al. (1983, 1985) produced an increase of approximately 50 percent in muscle glycogen concentration, above training effects, by feeding a low-carbohydrate diet during exhaustive, high-intensity exercise for 5 days to effect glycogen depletion, followed by a high-carbohydrate diet during a 3-day repletion period. Also, Pagan et al. (1987a) reported that muscle glycogen loading appeared to be most successful when high-carbohydrate diets were fed following intensive exercise bouts, which resulted in depletion of muscle glycogen stores. Simply increasing the carbohydrate concentration in the diet without simultaneously adjusting the training regimen may not alter muscle glycogen stores. Topliff et al. (1987) fed diets containing 15, 25, 33, or 42 percent starch to horses in training (galloping) for 16 weeks and found no difference in muscle glycogen concentrations at 0, 2, 4, 8, 12, or 24 h postfeeding.

Although increased athletic performance in the human is frequently associated with increased glycogen stores, similar effects have not been conclusively demonstrated in the horse. Several researchers have found that muscle glycogen stores in horses are not depleted to zero even after exhaustive exercise (Lindholm et al., 1974; Topliff et al., 1983, 1985; Miller et al., 1985). This has been interpreted to mean that glycogen depletion is not related to the onset of fatigue in horses (Snow and Vogel, 1987). However, Topliff et al. (1983, 1985) found that the onset of fatigue occurred sooner, and at lower heart rate and blood lactate concentration, when horses were depleted to muscle glycogen concentrations lower than

10 mg/g of wet tissue prior to long-term (30 min) or short-term (20 s) exhaustive exercise. Further, the amount of muscle glycogen mobilized during those exhaustive work bouts increased threefold when the preexercise muscle glycogen concentrations were increased from approximately 20 to 30 mg/g of wet tissue. Bayly et al. (1983) found that Thoroughbreds running a mile became oxygen deficient before the end of the mile; Webb et al. (1987) found that cutting horses used anaerobic energy production systems for most of their work. These studies suggest that muscle glycogen concentrations inadequate to facilitate sufficient anaerobic energy production for even a few seconds may limit performance of racehorses or other horses that perform short-term, high-velocity exercises. Snow et al. (1981) also concluded that horses performing endurance events were most likely to be successful when they had adequate muscle glycogen stores. However, Pagan et al. (1987b) expressed concern that horses with increased glycogen stores may have a greater tendency to develop exertional myopathy ("tying-up syndrome"). Further studies are required to clarify this relationship.

LIPIDS

Lipids are concentrated sources of energy that can be readily utilized by the horse. Body lipids are mobilized during exercise, and free fatty acids are oxidized readily, particularly during strenuous exercise such as galloping (Goodman et al., 1973; Anderson, 1975). However, these workers found that unconditioned horses do not appear to oxidize fat as efficiently as conditioned horses.

Hyperlipidemia is an important clinical problem in small pony breeds (Jeffcot and Field, 1985). It is most common in mares in late gestation and lactation, and occurs when the animal is in negative energy balance because of inappetence, starvation, pregnancy, lactation, parasitism, or transportation. Common clinical signs include dullness and lethargy, progressing to severe depression, coma, and death. The plasma or serum is opaque and milky looking, and serum triglycerides may increase to 500 mg/dl. Triglycerides in adipose tissue are hydrolyzed to fatty acids and glycerol. Most of the fatty acids are metabolized by the liver to yield energy and ketone bodies, or they can be reesterified to triglycerides. Fatty infiltration of the liver occurs in affected ponies, with concurrent hepatic dysfunction. Peripheral removal of lipids from plasma is inhibited, and lipid accumulates in the blood.

Many hyperlipidemic ponies also exhibit pancreatitis and reduced insulin production. Ponies, particularly when obese, may also be relatively insensitive to insulin, compared to horses. Jeffcott and Field (1985) proposed that this tissue insulin insensitivity exacerbates the effects of stress-induced cortisol release in the pathogenesis of both hyperlipidemia and laminitis.

Horses accept fat additions to the diet quite readily if the fat is not rancid. Bowman et al. (1979) compared the palatability of 10 fats or fat mixtures at 15 percent of the diet and found that horses seemed to prefer corn oil. Scott et al. (1987) found that additions of 5 or 10 percent fat to concentrates fed to geldings resulted in weight gains comparable to those of geldings on the unsupplemented diet. Feed intake tended to decline with increasing fat percentage, but diet acceptance was good. This decreased intake of higher energy density diets can be expected if horses, like many other species, eat to fulfill their energy requirements. Davison et al. (1987) added 5 percent feed-grade, rendered fat to concentrates fed to pregnant and lactating Quarter Horse mares and reported less feed consumption during gestation and a tendency toward an early growth advantage for foals, compared to mares and foals fed the unsupplemented diet. Inclusion of fat in the diet increased milk fat concentration at days 10 and 60 of lactation, but the percentages of protein and total solids were unaffected. Follicular activity during the transitional period just prior to the breeding season was not affected by feeding fat to mares in good body condition (Morris et al., 1987).

The effect of increasing dietary fat concentration on exercise performance has been studied by a number of researchers. Slade et al. (1975) reported that endurance horses fed a diet containing 12 percent fat (9 percent added corn oil) performed better and had higher blood glucose levels at the end of the ride than horses fed a diet with 3 percent fat. Hintz et al. (1978, 1982) included 8 percent animal fat in a diet containing corn and alfalfa meal in proportions of 60:40 and found a smaller decline in blood glucose of Thoroughbreds or Arabians ridden 60 km at 10 km/h or 83 km at 15 km/h, respectively, compared to controls fed no animal fat. Hambleton et al. (1980) fed diets containing 4, 8, 12, and 16 percent fat (as soybean oil) and found that postexercise plasma glucose concentrations correlated positively with dietary fat levels. Wolter et al. (1986) reported that the addition of fat improved the endurance capabilities of ponies. Meyers et al. (1987) reported that addition of 5 or 10 percent feed-grade, rendered animal fat to the concentrate fed to exercising horses spared muscle glycogen reserves. When Webb et al. (1987) added 10 percent fat to the concentrate fed to cutting horses, a higher blood lactate concentration at 30 s and 5 min postexercise, as well as a higher number of hindquarter turns during a cutting performance test, occurred than for cutting horses not fed supplemental fat. They interpreted these findings as evidence of an improved work performance compared to the performance of the same horses fed a concentrate without added fat.

Duren et al. (1987) fed Thoroughbred horses diets that provided 0, 5, 10, or 20 percent of their digestible energy concentration as corn oil. When exercised strenuously every other day on a work day–rest day rotation schedule, these horses used corn oil efficiently at all dietary levels in satisfying energy requirements. Fat had no negative effects on the items measured, and the declines in blood glucose concentration often associated with exercise diminished. However, when Pagan et al. (1987a) fed haylage plus a concentrate containing 15 percent soybean oil to Standardbred horses, muscle and liver glycogen decreased compared to horses fed high-carbohydrate diets subsequent to 31.5 km of submaximal exercise. Although this study and others demonstrate that the proportions of energy generated from fat and carbohydrate can be altered in exercising horses by dietary manipulation, the ideal proportions of these items in the diet for various intensities and durations of exercise remain to be determined.

Bowman et al. (1977) added 0, 5, 10, or 20 percent corn oil to a basal diet for ponies of chopped alfalfa hay, cracked corn, and crimped oats. Serum cholesterol concentrations increased due to the addition of corn oil; the values were 122, 144, 148, and 155 mg/dl, respectively. Serum triglyceride levels were unaffected. The apparent digestibility of the corn oil was 94 percent when it was added at 20 percent of the diet. The apparent digestibility of dietary protein was unaffected by dietary fat level. Corn oil added at 8 or 12 percent to a diet for ponies had a mean apparent digestibility of 93 percent, and there was no inhibition of dry matter or acid detergent fiber digestibility (Kane et al., 1979).

Although insufficient essential fatty acids (linoleic, γ-linolenic, or arachidonic acid) in the diet have been associated with hair and skin health in other species, the essential fatty acid requirements of the horse are not yet known. Regular grooming will improve the appearance of the horse. Until more data are available, dietary dry matter should include at least 0.5 percent linoleic acid.

Requirements

The energy requirements are expressed as digestible energy (DE). If the use of total digestible nutrients (TDN) is preferred, 4.4 Mcal of DE is assumed to be equivalent to 1 kg of TDN. Other systems such as metabolizable energy (ME) or net energy (NE) might be more precise. For example, an NE system based on the relative value of barley has been developed in France (Vermorel et al., 1984). Few ME or NE values for horses have been determined for U.S. feedstuffs; therefore, DE is used because more data are available. Energy values, such as TDN, DE, or NE, for feeds obtained from experiments with cattle may be significantly higher than those obtained from experiments with horses. Therefore, such energy values should not be used directly in the formulation of horse diets, particularly if the feed contains a significant amount of fiber.

Numerous factors such as individuality, body composition of the animal, environmental temperature and humidity, intensity and duration of work, weight and ability of the rider, conditions of the running surface, and degree of fatigue can influence the energy requirements of a horse. Therefore, the following amounts of energy should be considered as general guidelines.

MAINTENANCE

Values for energy requirements are not based on metabolic body size, because Pagan and Hintz (1986a) found no benefit from using metabolic body weight ($kg^{0.75}$) over weight ($kg^{1.0}$) in determining the energy requirements of horses ranging in size from 125 to 856 kg. The equation determined by Pagan and Hintz (1986a) for horses confined to metabolism stalls is

$$DE(Mcal/day) = 0.975 + 0.021W,$$

where W is the weight of the horse (kg).

Maintenance requirements (defined as the amount of DE required for zero body weight change plus normal activity of nonworking horses) of horses weighing 600 kg or less were estimated from the equation

$$DE(Mcal/day) = 1.4 + 0.03W,$$

where W is the weight of the horse (kg) (adapted from Pagan and Hintz, 1986a). The activity factor included in the above equation appears to overestimate the energy needs of horses of larger size because of their reduced voluntary activity (Pagan and Hintz, 1986a; Potter et al., 1987). Therefore, the digestible energy requirements for maintenance of horses with mature weights of more than 600 kg were estimated by the following:

$$DE(Mcal/day) = 1.82 + 0.0383W - 0.000015W^2,$$

where W is the weight of the horse (kg).

Values in Table 5-1 are consistent with other reports of energy requirements for maintenance (Brody, 1945; Morrison, 1956; Stillions and Nelson, 1972; Willard et al., 1978; Ellis and Lawrence, 1979; Jackson and Baker, 1981; Anderson et al., 1983).

REPRODUCTION

Breeding The body condition of mares at the time of breeding can influence the rate of conception. Henneke et al. (1984) reported that thin mares gaining weight at the time of breeding were twice as likely to conceive as thin mares maintaining weight. Mares in good to fat condition, however, had high conception rates even

when losing or maintaining body weight. Thus, the dietary energy needs for reproduction depend on the body condition of the mare. Henneke et al. (1983) developed the body score system shown in Table 2-1 and recommended that Quarter Horse mares at breeding time should have a minimum score of 5. Increasing the energy intake 10 to 15 percent above the requirement should cause weight gain and result in a higher condition score. Conversely, reducing the energy intake 10 to

TABLE 2-1 Description of Individual Condition Scores

Score	Description
1 Poor	Animal extremely emaciated; spinous processes, ribs, tailhead, tuber coxae, and ischii projecting prominently; bone structure of withers, shoulders, and neck easily noticeable; no fatty tissue can be felt
2 Very thin	Animal emaciated; slight fat covering over base of spinous processes; transverse processes of lumbar vertebrae feel rounded; spinous processes, ribs, tailhead, tuber coxae, and ischii prominent; withers, shoulders, and neck structure faintly discernible
3 Thin	Fat buildup about halfway on spinous processes; transverse processes cannot be felt; slight fat cover over ribs; spinous processes and ribs easily discernible; tailhead prominent, but individual vertebrae cannot be identified visually; tuber coxae appear rounded but easily discernible; tuber ischii not distinguishable; withers, shoulders, and neck accentuated
4 Moderately thin	Slight ridge along back; faint outline of ribs discernible; tailhead prominence depends on conformation, fat can be felt around it; tuber coxae not discernible; withers, shoulders, and neck not obviously thin
5 Moderate	Back is flat (no crease or ridge); ribs not visually distinguishable but easily felt; fat around tailhead beginning to feel spongy; withers appear rounded over spinous processes; shoulders and neck blend smoothly into body
6 Moderately fleshy	May have slight crease down back; fat over ribs spongy; fat around tailhead soft; fat beginning to be deposited along the side of withers, behind shoulders, and along the sides of neck
7 Fleshy	May have crease down back; individual ribs can be felt, but noticeable filling between ribs with fat; fat around tailhead soft; fat deposited along withers, behind shoulders, and along neck
8 Fat	Crease down back; difficult to feel ribs; fat around tailhead very soft; area along withers filled with fat; area behind shoulder filled with fat; noticeable thickening of neck; fat deposited along inner thighs
9 Extremely fat	Obvious crease down back; patchy fat appearing over ribs; bulging fat around tailhead, along withers, behind shoulders, and along neck; fat along inner thighs may rub together; flank filled with fat

SOURCE: Adapted from Henneke et al. (1983).

15 percent below the requirement should lead to weight loss and a lower score for the animal (Ott and Asquith, 1981).

Gestation The DE requirements for fetal development are not greatly increased until the last 3 months of gestation when the greatest development of the fetus occurs (Meyer and Ahlswede, 1976; Platt, 1984). Estimates of DE requirements for the ninth, tenth, and eleventh months of gestation were formulated by multiplying the maintenance requirements by 1.11, 1.13, and 1.20, respectively, and are similar to calculations by Meyer and Ahlswede (1976).

Lactation The DE requirements of lactating mares depend upon the composition and amount of milk produced. Studies of the effects of diet on composition of equine milk are limited. Baker and Jackson (1983) reported that the milk composition of mares fed hay and grain did not differ from those fed just hay. Pagan and Hintz (1986b) reported that increasing the energy intake of lactating pony mares to 20 percent above National Research Council (NRC, 1978) recommendations decreased the fat concentration of milk, but not total energy production.

Mares of light breeds appear to produce amounts of milk equivalent to 3 percent of body weight/day during early lactation (1–12 weeks) and 2 percent of body weight during late lactation (13–24 weeks) (Doreau et al., 1980; Gibbs et al., 1982; Oftedal et al., 1983). The milk production of ponies has been estimated to average 4 and 3 percent of body weight during early and late lactation, respectively (Neuhaus, 1959).

Zimmerman (1981) did not measure milk production, but from weight gains of foals and weight changes in mares, he concluded that the NRC (1978) may have slightly underestimated the energy requirements of lactating Quarter Horse mares. Estimates by Drepper et al. (1982) for light horse mares were similar to those of the NRC (1978). Pagan et al. (1984) and Jordan (1979), using criteria similar to those of Zimmerman (1981), suggested that estimates made in 1978 by the NRC for lactating pony mares, although slightly high, were reasonable. Therefore, the same assumptions used by the NRC (1978) for the conversion of DE into milk—that is, 792 kcal of DE/kg of milk—are used in this edition.

GROWTH

The following equations were developed from values reported in several sources (Householder et al., 1977; Milligan et al., 1985; Ott and Asquith, 1986; Schryver et al., 1987; Scott et al., 1987). Energy requirements for

TABLE 2-2 Digestible Energy Requirements for Growth of Foals

Age	Requirement (Mcal DE/kg of gain)
Weanling, 4 months	9.1
Weanling, 6 months	11.0
Yearling	15.5
Long yearling	18.4
2 years old	19.6

maintenance and growth rate were calculated by using the following equation:

$$DE(Mcal/day) = maintenance\ DE(Mcal/day) + (4.81 + 1.17X - 0.023X^2)(ADG),$$

(SE = 0.50, R^2 = 0.99), where ADG is the average daily gain (kg), and X is the age (months). The digestible energy requirement per kilogram of gain (Mcal/day), calculated as $4.81 + 1.17X - 0.023X^2$, increases with the age of the foal, as listed in Table 2-2.

The optimum growth rate and energy intake for productivity and longevity have not been established. High intakes of soluble carbohydrate have been associated with developmental orthopedic disease (osteochondrosis, epiphysitis, flexural deformities). Starch feeding caused significant changes in serum insulin, thyroxine, and triiodothyronine, which retarded cartilage maturation (Glade and Belling, 1984; Glade and Reimers, 1985; Glade and Luba, 1987). Thompson et al. (1987) concluded that a link exists between above-average weight gains and the onset of bone abnormalities, such as epiphysitis. They suggested that excessive force on the bone from heavy body weights can impair normal maturation of cartilage. However, Ott and Asquith (1986), Schryver et al. (1987), and Scott et al. (1987) found no orthopedic problems in rapidly growing foals.

Of course, several nutritional factors in addition to energy, such as vitamins and minerals, have been associated with developmental orthopedic disease (Gabel et al., 1988; Glade, 1988; Knight et al., 1988). Environmental influences other than nutrition, as well as genetic influences, may play a role in the etiology of orthopedic disease (Schougaard et al., 1987; Hintz, 1988; Kronfeld and Donoghue, 1988).

Table 5-1 illustrates the increased requirements for rapid growth.

WORK

Estimations of DE requirements for work are complicated because many factors that are difficult to quantitate can influence these requirements. For example, condition and training of the animals, ability and weight of the rider, degree of fatigue, environmental temperature, and diet composition may all have an effect.

The DE requirements for work have been estimated in several ways. For example, Anderson et al. (1983) suggested that the amount of DE needed for work and maintenance is best described by the quadratic equation:

$$DE(Mcal/day) = 5.97 + 0.021W + 5.03X - 0.48X^2,$$

where W is the body weight (kg), and $X = Z \times km \times 10^{-3}$

(Z = weight of horse, rider, and tack in kg).

The authors pointed out that this equation should not be used when the work load (kg × km) is greater than 3,560, which was the largest experimental value used in computing the equation. Therefore, the equation is particularly effective for horses performing intense work but not for horses performing endurance work (Ralston, 1988).

Pagan and Hintz (1986c) suggested that the DE required above maintenance (kcal/kg of weight of horse, rider, and tack/h) could be calculated by the equation

$$DE (kcal/kg/h) = \frac{[e^{(3.02 + 0.0065Y)} - 13.92] \times 0.06}{0.57},$$

where Y is the speed (m/min).

The equation of Pagan and Hintz (1986c) should not be used to estimate requirements of horses going faster than 350 m/min. Therefore, the equation is most effective estimating the energy requirements of horses performing for long periods of time.

A third method of estimating DE is to generalize according to activity. For example, data from surveys indicate that Thoroughbreds and Standardbreds in training or on the track consume an amount of feed equivalent to 2 to 3 percent of their body weight (Mullen et al., 1979; Nash and Hintz, 1981; Winter and Hintz, 1981; Glade, 1983; Burton and MacNeil, 1985; Minieri et al., 1985; Smith, 1986). Estimates of DE requirements from such data are influenced greatly by estimates of DE concentrations in the feed and estimates of feed intake.

After all the reports were reviewed, it was assumed that for ponies and light horses (200–600 kg), light, medium, and intense work as listed in Table 5-1 increased the energy requirements 25, 50, and 100 percent above maintenance, respectively.

The energy requirements for draft horses depend on many factors such as size of load and type of work. However, increasing maintenance requirements by 10 percent for each hour of field work should provide a reasonable guide (Brody, 1945).

PROTEIN

Protein accounts for about 22 percent of the fat-free composition of the mature horse and is 80 percent of the animal's structure on a fat-free, moisture-free basis (Robb et al., 1972). Amino acids, the building blocks of protein, are the major components of muscle, enzymes, and many hormones.

The dietary protein requirement of the horse is a function of the needs of the animal, the quality of protein available, and the digestibility of that protein. The influence of protein quality, or the relationship between the amino acid concentrations of the diet and the amino acid needs of the animal, on the growth of horses has been demonstrated by several studies. Weanling foals supplemented with linseed meal grew slower than foals fed milk protein (Hintz el al.,1969, 1971), and zein protein was found to be inferior to soybean meal protein (Breuer et al., 1970). Foals weaned early and fed dried skimmed milk grew faster than those fed diets based on soybean meal (Borton et al., 1973). Yearlings fed cottonseed meal (Potter and Huchton, 1975) or brewers dried grains (Ott et al., 1979) grew more slowly than those fed soybean meal. In all cases, addition of lysine to the low-quality protein resulted in gains comparable to those achieved on the higher-quality proteins, which suggests that the amino acid composition of the diet is critical for the growing horse.

Mature horses at maintenance are less sensitive to protein quality than growing horses. Corn gluten meal has been shown to be inferior to fish meal (Slade et al., 1970) or casein (Reitnour and Salsbury, 1976) as a protein source for maintenance, but some of the difference is probably due to digestibility. Protein digestibility varies with the source of the protein (Slade et al., 1970; Reitnour and Treece, 1971; Reitnour and Salsbury, 1976; Gibbs et al., 1988) and the concentration in the diet (Slade et al., 1970; Prior et al., 1974; Freeman et al., 1985; Gill et al., 1985). Digestible protein (DP) concentrations of various types of diets for mature horses can be estimated by using the following equations:

Grass hays*   DP(%) = 0.74 CP - 2.5
Oat hay:concentrate (1:1)†   DP(%) = 0.80 CP - 3.3
Alfalfa hay:concentrate (1:1)* DP(%) = 0.95 CP - 4.2

where CP is the crude protein concentration (percent).

From these equations the apparent protein digestibility may be as low as 43 percent on a low-protein (8 percent CP) grass hay and as high as 69 percent on a high-protein alfalfa-concentrate combination (16 percent CP). Increasing the concentrate-to-hay ratio above 1:1

---

*Fonnesbeck, P. V., Utah State University (unpublished data).
†Slade et al. (1970).

can result in higher protein digestibility (70 to 75 percent), probably because of an increased contribution from readily digestible ingredients such as soybean meal (Scott et al., 1987; Webb et al., 1987). However, very high concentrate-to-hay ratios (3:1) may reduce digestibility (Meyers et al., 1987). Limited information is available on the protein digestibility of individual feed ingredients. Because amino acids are absorbed predominantly from the small intestine and the predominant form of nitrogen absorbed from the hind gut is ammonia, the site of absorption also greatly influences the efficiency of protein utilization. Therefore, diet formulations for horses should be based on crude protein. When requirements were calculated from DP values, the protein digestibility estimates were based on the information cited above.

Requirements

MAINTENANCE

Estimates of the DP requirements for maintenance of horses have ranged from 0.49 to 0.68 g of DP/kg of body weight/day (Teeter et al., 1967; Slade et al.,1970; Hintz and Schryver, 1972; Prior et al., 1974; Quinn, 1975; Meyer et al., 1985). A value of 0.60 g of DP/kg of body weight/day appears to be appropriate for most horses. If the horse consumes a forage diet with a digestibility of 46 percent, the CP requirement would be 1.3 g of CP/kg of body weight/day. Dividing the crude protein requirement by the digestible energy requirements results in a value of 40 g of crude protein/Mcal of DE. The amino acid requirements of the mature horse have not been studied; however, when typical ingredients are used to meet the protein needs of the animal, adequate levels of essential amino acids apparently are provided. Protein in typical diets contains about 3.5 percent lysine.

REPRODUCTION

Breeding and Gestation   The protein requirements for nonlactating mares during the breeding season and early gestation have not been shown to be different from maintenance; however, Holtan and Hunt (1983) demonstrated that dietary protein concentrations influence circulating progesterone values and, therefore, have implications for mare fertility. Because 60 to 65 percent of fetal development occurs during the last 90 days of gestation (Bergin et al., 1967; Meyer and Ahlswede, 1978; Platt, 1984), protein requirements increase during this period.

Based on the fetal composition data of Meyer and Ahlswede (1978) and a protein utilization efficiency of 60 percent, a 500-kg gestating mare requires 127, 130, and 178 g of DP/day for fetal deposition above maintenance, for a total of 427, 430, and 478 g of DP during the ninth, tenth, and eleventh months, respectively. If 55 percent protein digestibility is assumed, the 500-kg mare would need 776, 782, and 869 g of CP daily during the ninth, tenth, and eleventh months, respectively. This represents a protein-energy ratio of approximately 44 g of CP/Mcal of DE, which was used to calculate the crude protein requirements in Table 5-1.

Lactation   The protein content of milk (Appendix Table 1) is highest immediately after parturition and decreases gradually as lactation progresses (Ullrey et al., 1966; Gibbs et al., 1982; Oftedal et al., 1983; Zimmerman, 1985). Milk composition does not appear to be influenced by diet (Zimmerman, 1985), but the level of production depends on energy and protein intake (Jordan, 1979; Banach and Evans, 1981).

Using means of the production figures discussed in the energy section (lactation), CP requirements were calculated assuming the following: (1) mares' milk contains 2.1 and 1.8 percent protein in early and late lactation, respectively (Appendix Table 1); (2) DP is used with an efficiency of 65 percent for milk production; and (3) protein digestibility in typical lactation diets is 55 percent. The daily protein requirement of the 500-kg lactating mare during early lactation would be calculated as follows:

500 kg × 0.03 kg of milk produced per kg of BW × 2.1 percent protein in milk = 0.315 kg of protein secretion

0.315 kg/efficiency of utilization coefficient (0.65) = 0.485 kg of protein for milk production

0.485 kg + maintenance requirement (0.3 kg) = 0.785 kg of dietary DP required

0.785 kg/apparent digestibility coefficient (0.55) = 1.427 kg of dietary CP required

The resulting recommendations are 5 percent above the 1978 recommendation for early lactation and 5 percent below the 1978 recommendation for late lactation, but are compatible with the work of Gill et al. (1985), which suggested that the 1978 recommendations were too low for maximum milk production early in lactation.

GROWTH

Protein and energy intakes are the major nutrient factors influencing the growth of young horses. Restricting either protein intake (Ott et al., 1979; Meakim et al., 1981) or energy intake of the growing horse (Ellis and

Lawrence, 1979; Ott and Asquith, 1986) restricts growth rate. Therefore, there is an optimal relationship between energy and protein intake for the growing horse. Based on available information, CP requirements were calculated to be 50 and 45 g/Mcal of DE/day for weanlings and yearlings, respectively (Wirth, 1977; Meakim, 1979; Ott et al., 1979; Potter, 1981; Ott and Asquith, 1983, 1986; Boren et al., 1987; Gibbs et al., 1987; Scott et al., 1987; Szcurek et al., 1987).

Lysine is the first limiting amino acid in the diet of growing foals (Breuer et al., 1970; Breuer and Golden, 1971; Hintz et al., 1971; Potter and Huchton, 1975; Ott et al., 1979, 1981; Ott and Asquith, 1986). Calculations from the studies cited indicate that the lysine requirements are 2.1 and 1.9 g/Mcal of DE/day for weanlings and yearlings, respectively. Data on other amino acids are inadequate to permit a recommendation.

WORK

The effect of exercise on the protein requirement of the horse has not been elucidated adequately. Although Harvey et al. (1939) and Patterson et al. (1985) suggested that exercise has little or no effect on the protein requirement of the horse, Freeman et al. (1988) documented increased apparent nitrogen retention in working horses which included unmeasured losses in the sweat. This nitrogen retention includes muscle hypertrophy as animals become physically conditioned (Freeman et al., 1988), and perhaps increased muscle protein content (Freeman et al., 1985) and nitrogen lost via sweat. Sweat will contain 1–1.5 g of nitrogen/kg (Meyer, 1987). Sweat loss increases as activity levels increase, and losses as high as 5 kg/100 kg of body weight have been estimated (Meyer, 1987). At a dietary DP:DE ratio sufficient for maintenance, the increased DE intake necessary to maintain body condition in a working horse will provide adequate additional nitrogen to meet the working horse's requirements (Hinkle et al., 1981; Freeman et al., 1988). Feeding high-protein diets to working horses is not advantageous. Hinkle et al. (1981) found similar nitrogen retention for working horses consuming 0.93 to 1.19 kg of CP/day.

Utilization of Nonprotein Nitrogen

There appears to be no beneficial effect of including nonprotein nitrogen (NPN) sources in practical diets for horses. Urea is well tolerated by the mature horse at levels up to 4 percent of the total diet. This tolerance is likely due to the high solubility of urea, which results in its absorption from the small intestine and subsequent excretion by the kidney (Slade et al., 1970; Nelson and Tyznik, 1971; Reitnour and Treece, 1971; Hintz and Schryver, 1972; Reitnour, 1978). Urea is excreted into the hind gut (Nelson and Tyznik, 1971; Prior et al., 1974) when blood urea levels are high, which may enhance microbial metabolism. Some workers have reported increased nitrogen retention by horses fed nonprotein sources (Slade et al., 1970; Hintz and Schryver, 1972), whereas others have shown no increased retention (Nelson and Tyznik, 1971; Reitnour and Treece, 1971; Quinn, 1975; Reitnour, 1978). Mature horses consuming low-nitrogen diets appear to be able to utilize some urea nitrogen to meet daily nitrogen requirements. Growing horses cannot achieve maximum growth rates (Godbee and Slade, 1981), and lactating mares do not give maximum milk production when a major portion of the nitrogen requirement is provided by urea (Gibbs et al., 1982).

Although ponies succumb to single doses of 0.5 kg of urea (Hintz et al., 1970), intakes of feed containing up to 5 percent urea and providing as much as 0.25 kg of urea daily did not have any detrimental effects on mature horses (Ratliff et al., 1963; Rusoff et al., 1965).

Signs of Deficiency or Excess

Inadequate protein or lysine intake results in decreased growth and development of young animals. Inadequate intake of protein or required amino acids by mature horses may lead to reduced feed intake, body tissue loss, poor hair coat, and reduced hoof growth.

Moderate excesses of protein have no detrimental effect on horses. Protein intakes of 64 g of CP/Mcal of DE by weanlings (Yoakam et al., 1978) and 56 g of CP/Mcal of DE by yearlings (Gibbs et al., 1987) resulted in growth depression. These intakes are at least 25 percent above the requirement. Meakim et al. (1981), however, found no adverse effects on growth or calcium metabolism in weanlings fed 59 g of CP/Mcal of DE. Glade et al. (1985), in a short-term metabolism study, found evidence that protein intakes similar to those cited above cause a reduction in renal calcium reabsorption.

Although Slade et al. (1975) suggested that protein intake considerably above the requirement is detrimental to working horses, others (Hintz et al., 1980; Ralston, 1988) were unable to find any beneficial or detrimental effects from feeding high levels of protein to endurance horses, except that the daily water requirements increased. Glade (1983) suggested from a study of the feeding programs of seven Thoroughbred trainers that excess protein intake may adversely affect performance. However, that study was confounded by the observation that high energy intake also correlated with reduced performance; therefore, no cause and effect can be established from such studies.

MINERALS

Minerals are involved in a number of functions in the body, including formation of structural components, enzymatic cofactors, and energy transfer. Some minerals are integral parts of vitamins, hormones, and amino acids. The horse obtains most of the necessary minerals from pasture, roughage, and grain. The mineral content of feeds and the availability of minerals vary with soil mineral concentrations, plant species, stage of maturity, and conditions of harvesting. The resulting variations should be considered in assessing an animal's mineral status and formulating appropriate diets.

Seven macrominerals (calcium, phosphorus, potassium, sodium, chloride, magnesium, sulfur) and eight microminerals (cobalt, copper, fluorine, iodine, iron, manganese, selenium, zinc) are discussed below.

Macrominerals

CALCIUM

Calcium makes up about 35 percent of bone structure (El Shorafa et al., 1979) and is involved in other body functions including muscle contractions and blood clotting mechanisms. The true absorption efficiency is believed to decline with age and to range from as high as 70 percent in young horses to 50 percent in mature horses. For purposes of estimating calcium requirements herein, a calcium absorption efficiency of 50 percent is used for all ages of horses.

Other factors affecting calcium absorption include the dietary concentrations of calcium, phosphorus, oxalate, and phytate. Swartzmann et al. (1978) reported that 1 percent oxalic acid in equine diets reduced calcium absorption approximately 66 percent. Total dietary oxalate concentrations of 2.6 to 4.3 percent produced negative calcium balance, during which fecal calcium doubled and urinary calcium decreased in comparison to control horses (McKenzie et al., 1981). Blaney et al. (1981) observed similar negative balances for calcium and phosphorus in horses fed various tropical grass hays of more than 0.5 percent total oxalate. They concluded that nutritional secondary hyperparathyroidism may occur when the calcium-oxalate ratio on a weight-to-weight basis is less than 0.5. Hintz et al. (1984) reported no difference in absorption of calcium from alfalfas containing 0.5 and 0.87 percent oxalic acid, in which the calcium:oxalate ratios were 3 and 1.7, respectively.

Estimates of endogenous losses of 20 mg of calcium/kg of body weight/day have been made (Schryver et al., 1970, 1971a). If the absorption efficiency of dietary calcium is 50 percent, a 500-kg horse would require 20 g (500 kg × 20 mg/0.5) of dietary calcium or 0.04 g of calcium/kg of body weight/day or 1.22 g of calcium/Mcal of DE/day for maintenance. Schryver et al. (1974) estimated that growing foals deposit approximately 16 g of calcium/kg of gain. Thus, a 215-kg foal gaining 0.85 kg/day and having a calcium absorption efficiency of 50 percent would require 27.2 g/day (16 g × 0.85 kg/0.5) of dietary calcium for skeletal growth plus 8.6 g/day (215 kg × 20 mg/0.5) to meet endogenous losses.

Any increase in the calcium requirement associated with exercise (work) appears to be readily met by an obligatory increase in calcium intake as dry matter consumption increases to meet energy requirements. Special calcium requirements occur in late gestation and during lactation. Approximately 11.1, 25.3, and 11.4 mg of calcium/kg of mare body weight are deposited daily in the fetus and membranes of mares in months 9, 10, and 11 of gestation, respectively (Meyer and Ahlswede, l976; Drepper et al. 1982). If the absorption efficiency is 50 percent, the calcium requirement of a 500-kg mare during months 9, 10, and 11 of gestation would be 11, 25, and 11 g/day, respectively (mean 15.9 g/day), for fetal development. Data on the deposition rate of minerals in the fetus are very limited; therefore, a mean deposition rate was used for the last 3 months. The mean dietary calcium required for tissue deposit was added to maintenance needs and divided by the mean DE requirement for the same period. This value was then multiplied by the daily DE requirement for each month.

Schryver et al. (1986) reported that the daily calcium requirements for lactation range from 1.2 g/kg of fluid milk during the first postpartum week to 0.8 g/kg of fluid milk during weeks 15 to 17 postpartum. This is consistent with the data of Baucus et al. (1987). For a 500-kg mare producing 15 kg of milk/day in early lactation and with a calcium absorption efficiency of 50 percent, the daily dietary calcium requirement would be 36 g (15 kg × 1.2 g/0.5) for milk production, in addition to 20 g for maintenance. This estimate of calcium requirement for milk production is consistent with that of Jarrige and Martin-Rosset (1981).

Calcium homeostatic mechanisms maintain serum calcium within a narrow range. Therefore, serum calcium may not be a reliable indicator of calcium status (Krook and Lowe, 1964). Caple et al. (1982) have proposed that adequate calcium intake exists in horses excreting more than 15 μmol calcium/mosm of urine solute when calcium:creatinine clearance ratio is greater than 2.5.

Signs of Deficiency or Excess   Inadequate calcium intake by the developing foal can lead to rickets, which is characterized by poor mineralization of the osteoid tissue and the probability of enlarged joints and crooked long bones. From a survey of the severity of metabolic bone disease in yearlings and diet analyses on 19 Ohio and Kentucky horse farms, Knight et al. (1985) reported a negative linear relationship between dietary calcium intake and perceived severity of metabolic bone disease in young horses. Farms with yearlings having the lowest incidence of metabolic bone disease were fed diets containing 1.2 percent calcium, whereas yearlings with the most severe metabolic bone disease were on farms that fed diets with 0.2 percent calcium. In the mature horse, inadequate dietary calcium can result in weakening of the bones and an insidious shifting lameness (Krook and Lowe, 1964).

Whitlock et al. (1970) fed diets with calcium:phosphorus ratios of 1.16:1 (0.43 percent calcium) and 4.12:1 (1.96 percent calcium) and observed a greater proportion of lamellar bone than osteonic bone in the high-calcium horses; however, no clinically deleterious effects or gross morphologic differences were detected. Krook and Maylin (1988) proposed that osteochondrosis may be associated with excess dietary calcium (e.g., from alfalfa hay) and its production of hypercalcitoninism. However, calcium has been fed at more than 5 times the required level without detrimental effects, provided the level of phosphorus is adequate (Jordan et al., 1975).

PHOSPHORUS

Phosphorus makes up 14–17 percent of the skeleton (El Shorafa et al., 1979). In addition, it is required for many energy transfer reactions associated with adenosine diphosphate (ADP) and adenosine triphosphate (ATP), and for the synthesis of phospholipids, nucleic acids, and phosphoproteins. The endogenous phosphorus loss in mature horses has been estimated at 10 mg/kg of body weight/day (Schryver et al., 1971b).

The efficiency of estimated true phosphorus absorption ranges from 30 to 55 percent and varies with the age of the animal and with the source and concentration of dietary phosphorus. Phosphorus absorption is likely to be higher in foals fed milk than in older horses. Phosphorus in the form of phytate is poorly absorbed; however, phytate phosphorus may be partially available because there is some phytase in the equine lower gut (Hintz et al., 1973). Phosphorus retention is depressed in natural diets containing high concentrations of total oxalates (Blaney et al., 1981; McKenzie et al., 1981). In this document, a true phosphorus absorption efficiency of 35 percent is used for all idle horses, gestating mares, and working horses, because they consume primarily plant sources of phosphorus; a value of 45 percent is used for lactating mares and growing horses because their diets are typically supplemented with inorganic phosphorus.

At a phosphorus absorption efficiency of 35 percent, the phosphorus requirement for maintenance is 10 mg/0.35 or 28.6 mg/kg of body weight/day or 0.87 g of phosphorus/Mcal of DE/day. A mature, 500-kg horse would, therefore, require 14.3 g of phosphorus/day for maintenance. The phosphorus requirements for optimal bone development in growing foals are based on estimates that growing horses deposit approximately 8 g of phosphorus/kg of body weight gain (Schryver et al., 1974). Thus, a 215-kg foal with a phosphorus absorption efficiency of 45 percent and gaining 0.85 kg/day would require about 15.1 g of phosphorus (8 g × 0.85 kg/0.45) in addition to its maintenance requirement of 4.8 g (215 kg × 10 mg/0.45).

During late gestation and lactation, phosphorus requirements increase. Phosphorus requirements for the products of conception for mares in months 9, 10, and 11 of pregnancy have been estimated to be 7, 12, and 6.7 mg/kg of body weight/day, respectively (Drepper et al., 1982). At 35 percent absorption efficiency, the total daily phosphorus requirement for a 500-kg mare during gestation months 9, 10, and 11 would be 7.8, 13, and 7.4 g, respectively (mean 9.4 g/day). Data on the rate of deposition of minerals in the fetus are very limited. Therefore, a mean deposition rate was used for the last 3 months. Mean dietary phosphorus required for tissue deposit was added to maintenance needs and divided by the mean DE requirement for the same period. This value (1.44 g of phosphorus/Mcal of DE/day) was then multiplied by the daily DE requirement for each month.

The phosphorus concentration of mares' milk ranges from 0.75 g/kg of fluid milk in early lactation to 0.50 g/kg of fluid milk in late lactation (Appendix Table 1). If the absorption efficiency is 45 percent, the daily phosphorus requirement above maintenance for lactation would be 25.0 g for a mare averaging 15 kg of milk/day in early lactation and 11.1 g for the mare producing 10 kg of milk/day during late lactation. At these rates of milk production, a 500-kg mare would require 36 and 22.2 g of phosphorus/day in early and late lactation, respectively.

Calcium-Phosphorus Ratio The calcium-phosphorus ratio is an important criterion for equine diets. Ratios less than 1:1 (i.e., when phosphorus intake exceeds calcium intake) may be detrimental to calcium absorption. Even if the calcium requirements are met, excessive phosphorus intake will cause skeletal malformations (Schryver et al., 1971b). Ratios of calcium to phosphorus as high as 6:1 in diets for growing horses may not be detrimental if phosphorus intake is adequate (Jordan et al., 1975).

Signs of Deficiency or Excess Inadequate dietary phosphorus will, like inadequate calcium and vitamin D, produce rachitic changes in growing horses and osteomalacic changes in mature horses. As indicated, excess phosphorus reduces the rate of calcium absorption and leads to chronic calcium deficiency and nutritional secondary hyperparathyroidism, also known as big head, miller's disease, bran disease, osteofibrosis, and fibrous osteodystrophy. Nutritional secondary hyperparathyroidism is characterized by a shifting lameness and, in advanced cases, by enlargement of the upper and lower jaws and facial crest (Krook and Lowe, 1964).

Serum inorganic phosphorus may be more indicative of dietary phosphorus status than serum calcium is of calcium status because the homeostatic mechanisms for phosphorus are less sensitive than for calcium (Schryver et al., 1970, 1971b). Caple et al. (1982) determined that horses excreting more than 15 μmol of phosphorus/mosm of urine solute and having a phosphorus:creatinine clearance ratio greater than 4 had excessive phosphorus intake and were subject to nutritional secondary hyperparathyroidism. The NRC (1980) suggested that a maximum tolerable level of dietary phosphorus in horses fed adequate dietary calcium is 1 percent.

POTASSIUM

Potassium is the major intracellular cation; it is involved in maintenance of acid-base balance and osmotic pressure. Forages and oilseed meals generally contain 1 to 2 percent potassium, in the dry matter, whereas cereal grains (corn, oats, wheat) contain 0.3 to 0.4 percent potassium. The required potassium concentration in a purified-type diet for growing foals was estimated at 1 percent (Stowe, 1971). Hintz and Schryver (1976) estimated that mature horses required 0.06 g of potassium/kg of body weight/day, or approximately 0.4 percent of the diet. Because forages usually constitute a significant portion of the diet, the potassium requirements for horses should be met easily. If necessary, potassium chloride and potassium carbonate are effective sources of supplemental potassium.

Jarrige and Martin-Rosset (1981) indicated that optimal potassium concentrations of equine diets were 0.4 to 0.5 percent for light to medium work, 0.4 percent for late gestation, 0.6 percent for 6- to 12-month-old foals, and 0.8 percent for horses 18 to 24 months of age. Drepper et al. (1982) indicated that the products of conception require 1.2, 1.7, and 2.2 mg of potassium/kg of mare weight during gestation months 9, 10, and 11, respectively. Further, Drepper et al. (1982) estimated the daily potassium requirements for a 600-kg horse to be

22 g for maintenance, 32 g for light work, 43 g for medium work, 53 g for heavy work, 22 g during the eighth to ninth months of gestation, 23 g for the tenth to eleventh months of gestation, and 34 g during lactation. For growth of foals with an anticipated body weight of 600 kg, the daily potassium requirements were estimated to be 11 g for months 3 to 6, 14 g for months 7 to 12, and 18 g for months 13 to 24.

Based on the research cited above, the daily potassium requirement for maintenance is estimated to be 0.05 g/kg of body weight or 1.52 g/Mcal of DE. Therefore, a 500-kg horse would require 25 g (500 kg × 0.05 g) of dietary potassium daily for maintenance. If this requirement is applied to the work of Drepper et al. (1982), the requirements for light, medium, and heavy work become 1.1, 1.4, and 1.8 times maintenance, respectively. This relationship can be approximated by the potassium to DE relationships given above. Growing horses have been shown to deposit 1.5 g of potassium/kg of gain (Schryver et al., 1974). Thus, a 215-kg foal gaining 0.85 kg/day and having a true potassium retention efficiency of 50 percent requires 2.6 g (1.5 g × 0.85 kg/0.5) of dietary potassium/day for skeletal growth in addition to 10.8 g (0.05 g × 215 kg) for maintenance.

Signs of Deficiency or Excess Foals fed potassium-deficient, pelleted, purified diets gradually refused to eat and, therefore, lost weight, became unthrifty in appearance, and had moderately lowered serum potassium concentrations (hypokalemia). On addition of potassium carbonate to the purified diet, an immediate resumption of normal feed intake occurred (Stowe, 1971). Excess dietary potassium is excreted readily, primarily via the urine, when water intake is adequate. The effects of excess potassium have not been studied in the horse; however, hyperkalemia, induced by parenteral administration of excess potassium, would be expected to cause cardiac arrest (Tasker, 1980).

SODIUM

Sodium is the major extracellular cation and the major electrolyte involved in maintenance of acid-base balance and osmotic regulation of body fluids. The sodium concentration of natural feedstuffs for horses is often lower than 0.1 percent. Sodium chloride (common salt) is often added to concentrates at rates of 0.5 to 1.0 percent or fed free-choice as plain, iodized, or trace-mineralized salt. Endogenous sodium loss in the idle adult horse has been estimated at 15 to 20 mg/kg of body weight/day (Meyer et al., 1984; Schryver et al., 1987). If sodium is 100 percent absorbed, then a 500-kg horse requires 7.5 g of sodium to meet the daily requirement. This amount would be provided by 19 to 25 g of sodium chloride/day.

Optimal sodium concentrations for equine diets have been reported to be between 1.6 and 1.8 g/kg of dry matter for growth, maintenance, and late gestation and 3.6 g/kg of dry matter for moderate to heavy work (Jarrige and Martin-Rosset, 1981). Drepper et al. (1982) reported that the sodium requirement of pregnant mares, above maintenance, is approximately 1.9 mg/kg of body weight/day during the tenth month of gestation. These authors also indicated that the sodium requirements (g/day) for the mature 600-kg horse are 15 for maintenance, 21 to 36 for light to heavy work, 16 for late gestation, and 18 for lactation. Prolonged exercise and elevated temperatures increase the sodium requirement because sweat contains a significant amount of sodium. Sweat sodium losses are estimated to range from 8.25 to 82.5 g (Meyer, 1987). Meyer et al. (1984) reported that a negative sodium balance could be demonstrated transiently in nonexercised horses and ponies after initial restriction of sodium intake to 5 mg/kg of body weight/ day. However, over time, these horses adapted to sodium restriction and, ultimately, could be in positive sodium balance while consuming only 1.6 mg of sodium/kg of body weight/day. Because of limited data on specific requirements for sodium and the influence of activity, adaptation, and environment on animal needs, precise recommendations cannot be made. However, sodium concentration in the maintenance diet should be at least 0.1 percent.

Signs of Deficiency   Chronic sodium depletion results in decreased skin turgor, a tendency for horses to lick objects such as sweat-contaminated tool handles, a slowed rate of eating, decreased water intake, and eventually a cessation of eating (Meyer et al., 1984). In acute sodium deficiency, muscle contractions and chewing are uncoordinated and horses have an unsteady gait; serum sodium and chloride concentrations decrease markedly, whereas serum potassium increases (Meyer et al., 1984).

CHLORIDE

Chloride, which normally accompanies sodium, is an important extracellular anion involved in acid–base balance and osmotic regulation. Chloride is an essential component of bile and is important in the formation of hydrochloric acid, a component of gastric secretions necessary for digestion. The chloride requirements of horses have not been specifically established. Chloride requirements are presumed to be adequate when the so-dium requirements are met with sodium chloride. Some chloride concentrations of common equine feedstuffs range from 0.05 percent for corn and soybean meal to 3 percent for molasses (NRC, 1982).

Signs of Deficiency or Excess   Chloride deficiency in horses has not been described; however, if a deficiency occurred, it should result in blood alkalosis because of a compensatory increase in bicarbonate during the chloride deficit (Tasker, 1980). Clinical signs of chloride deficiency may be similar to those reported in ruminants and include decreased food intake, weight loss, muscle weakness, decreased milk production, dehydration, constipation, and depraved appetite (Fettman et al., 1984).

Horses are considered tolerant of high levels of salt in their diets if they have free access to fresh drinking water. High salt concentrations in feeds are sometimes used to limit feed intake, especially of supplements. Parker (1984) reported that ponies consumed a 3-day grain ration over 3 days when the grain contained 16 percent salt but consumed the same ration in 1 to 2 days when the grain contained only 4 to 8 percent salt. The elevated dietary salt concentrations were associated with marked increases in water intake. Regulating the concentrate intake by salt addition is generally not as effective in horses as in ruminants. The maximum percentages of the daily salt requirements tolerated in drinking water for 450-kg working and lactating horses were estimated at 840 and 1,050 percent, respectively (NRC, 1974). Central nervous system manifestations of salt toxicity occur in some species, and horses can be expected to respond similarly.

MAGNESIUM

Magnesium constitutes approximately 0.05 percent of body mass, of which about 60 percent is associated with the skeleton. In addition, magnesium is an activator of many enzymes. Magnesium concentrations of common equine feedstuffs range from 0.1 to 0.3 percent. The true absorption of magnesium from feeds is in the range of 40 to 60 percent (Hintz and Schryver, 1972, 1973; Meyer, 1979). Supplemental dietary magnesium in sources such as magnesium oxide, magnesium sulfate, and magnesium carbonate is apparently absorbed at approximately 70 percent (Harrington and Walsh, 1980). McKenzie et al. (1981) reported magnesium to be 42 to 45 percent digestible and not significantly affected by oxalate. Endogenous excretion of magnesium by the mature horse is estimated at 6 mg/kg of body weight/ day.

If a true absorption efficiency of 40 percent is used for

magnesium, the daily magnesium requirement is approximately 15 mg/kg of body weight or 0.46 g/Mcal of DE; thus, a 500-kg horse at maintenance requires 7.5 g of dietary magnesium/day. This is much lower than the 12-g daily magnesium intake proposed by Drepper et al. (1982) for maintenance of a 600-kg horse. Schryver et al. (1974) determined that 0.85 to 1.25 g of magnesium is required/kg of body weight gain/day. From the latter figure, a 215-kg foal gaining 0.85 kg/day would need 1.07 g of magnesium for growth plus 3.23 g for endogenous fecal loss, for a total of 4.30 g/day. Drepper et al. (1982) suggested that light to medium work increases the magnesium requirement by 1 to 2 g/day for a 600-kg horse. Wolter et al. (1986) recommended supplementing 0.18 percent dietary magnesium for horses in training, especially if supplemental fat is fed.

The magnesium requirement of the mare associated with products of conception has been estimated at 0.23, 0.31, and 0.36 mg/kg of body weight of mare for months 9, 10, and 11, respectively (Drepper et al., 1982). To meet the magnesium requirement for these periods, a 500-kg mare would need 287, 387, and 450 mg of dietary magnesium daily. Data on deposition rate of minerals in the fetus are very limited; therefore, a mean deposition rate was used for the last 3 months. Mean dietary magnesium required for tissue deposit was added to maintenance needs and divided by the mean DE requirement for the same period. The resulting value (0.48 g of magnesium/Mcal of DE) was then multiplied by the daily DE requirement for each month.

The magnesium concentration of milk averages 90 μg/g during early lactation and 45 μg/g during late lactation (Appendix Table 1). If the absorption efficiency is 40 percent, a mare producing 15 kg of milk/day with a magnesium concentration of 90 μg/g would require 3.4 g of dietary magnesium/day for milk production in early lactation, in addition to the 7.5 g of magnesium required for maintenance.

Signs of Deficiency or Excess   Meyer and Ahlswede (1977) indicated that a magnesium intake of 5 to 6 mg/kg of body weight/day resulted in hypomagnesemia (less than 1.6 mg/dl) and marked reduction in renal excretion of magnesium, whereas 20 mg of magnesium/kg of body weight/day resulted in normal serum magnesium values of 1.6 to 2.0 mg/dl. Experimental production of hypomagnesemia in foals fed a purified diet containing 7 to 8 mg of magnesium/kg of diet for 150 to 190 days resulted in nervousness, muscle tremors, and ataxia, followed by collapse, hyperpnea, sweating, convulsive paddling, and some deaths (Harrington, 1974). As with other species, hypomagnesemia induces mineralization (focal calcium and phosphorus deposits) in the aorta. Histo-logic changes occur within 30 days of initiation of a low-magnesium diet (Harrington, 1974).

Pastures that are conducive to magnesium deficiency, tetany, and death in ruminants do not affect horses similarly. However, tetany in transported horses has been attributed to (or associated with) hypomagnesemia (Green et al., 1935).

There are no controlled studies on the effects of excess dietary magnesium for horses. Although the maximal tolerable level of dietary magnesium has been estimated as 0.3 percent from data on other species (NRC, 1980), some alfalfa hays with magnesium concentrations of 0.5 percent have been fed to horses without apparent ill effects (Lloyd et al., 1987). The source of magnesium may be important for horses since the use of dolomitic limestone in a concentrate containing 0.6 percent magnesium reduced the intake of this concentrate (G. Potter, Texas A&M University, personal communication, 1988), whereas no adverse effects were noted in mature ponies fed diets containing 0.86 percent magnesium for 1 month when the principal source was magnesium oxide (Hintz et al., 1973). Historically, magnesium sulfate was used intravenously as an anesthetic agent in horses prior to the advent of barbiturates and inhalation anesthetics (Kato et al., 1968). However, there is no indication that magnesium sulfate has an anesthetic effect at normal dietary concentrations.

SULFUR

Sulfur, in the form of sulfur-containing amino acids, biotin, heparin, thiamin, insulin, and chondroitin sulfate, makes up about 0.15 percent of the body weight. The sulfur requirements of the horse have not been established. Nonruminant animals must meet their sulfur amino acid requirements from organic sulfur forms, such as preformed cystine and methionine. Although some dietary inorganic sulfur is incorporated into sulfur-containing microbial protein in the equine hind gut, amino acid absorption from this region is limited (see protein section). Inorganic forms of dietary sulfur are used in the synthesis of some sulfur-containing substances such as chondroitin sulfate, heparin, and insulin. Adequate, high-quality dietary protein usually provides at least 0.15 percent organic sulfur. This appears adequate to meet the sulfur requirements of the horse (NRC, 1978; Jarrige and Martin-Rosset, 1981).

Signs of Deficiency or Excess   Sulfur deficiency in horses has not been described. Maximum tolerable dietary sulfur levels have not been estimated. Corke (1981), however, reported the effects of excess sulfur on 5- to 12-year-old horses that were accidentally fed between 200 and 400 g of flowers of sulfur (>99 percent sulfur). The horses became lethargic within 12 h, and colic often supervened. Other signs included a yellow, frothy discharge from the external nares, jaundiced mucous membranes, and labored breathing. Two of the twelve horses developed an expiratory snort, and cyanosis; despite treatment, they died following convulsions. Chronic consumption of excess sulfur in ruminants depresses copper absorption and can induce secondary copper deficiencies. There is no evidence that the equine species is subject to this effect of sulfur (Strickland et al., 1987).

Trace Minerals

COBALT

Cobalt is an integral part of the vitamin B_{12} molecule, whose function is discussed under vitamins. The cobalt requirements of horses have not been studied specifically. Cecal and colonic microflora of horses use dietary cobalt in the synthesis of vitamin B_{12} (Davies, 1971; Salminen, 1975). Filmer (1933) reported that horses remained in good health while grazing pastures that were inadequate in cobalt for ruminants.

COPPER

Copper is essential for several copper-dependent enzymes involved in the synthesis and maintenance of elastic connective tissue, the mobilization of iron stores, preservation of the integrity of mitochondria, melanin synthesis, and the detoxification of superoxide. The efficiency of copper absorption is inversely related to the dietary copper concentration (Cymbaluk et al., 1981a). The copper concentration of common equine feedstuffs ranges widely from approximately 4 mg/kg for corn to 80 mg/kg for cane molasses. Although no controlled studies on comparative copper availability have been reported for the horse, salts such as cupric chloride, cupric sulfate, and cupric carbonate are effective supplemental copper sources in other nonruminants (Cromwell et al., 1978, 1984).

Satisfactory growth was attained by foals fed 9 mg of copper/kg of diet (Cupps and Howell, 1949), whereas normal copper homeostasis was maintained in mature ponies fed 3.5 mg of copper/kg of diet (Cymbaluk et al., 1981a). Jarrige and Martin-Rosset (1981) and Drepper et al. (1982) recommended 10 mg of copper/kg of diet for all ages of horses, regardless of degree of work or stage of reproduction. Knight et al. (1985) reported a negative correlation between the copper concentrations of weanling diets and a perceived degree of affliction

with metabolic bone disease. In a subsequent study (Knight et al., 1988), mares were fed 13 and 32 mg of copper/kg of diet and their foals were fed a creep feed containing 15 and 55 mg of copper/kg. All foals were healthy, grew normally, and showed no signs of lameness or ataxia. Histologic lesions were reported for each group without statistical inference. Thus, the subcommittee considers the data to be inconclusive. Additional research on the copper requirements of the foal is necessary.

Several factors can influence copper metabolism. Copper interacts with many other minerals including molybdenum, sulfate, zinc, selenium, silver, cadmium, iron, and lead (Underwood, 1981). The quantitative relationships among these minerals have not been studied in detail in the horse; however, some distinct species differences are known. Molybdenum intakes at 1 to 3 mg/kg of diet interfere with copper utilization in ruminants (Underwood, 1981), but much higher levels of molybdenum are tolerated by the horse (Underwood, 1977; Cymbaluk et al., 1981b). Therefore, the likelihood of a molybdenum problem in the horse is minimal (Strickland et al., 1987). The effect of zinc on copper metabolism is discussed in more detail under zinc.

Signs of Deficiency or Excess There are reports that osteochondrosis and osteodysgenesis are associated with hypocupremia (Bridges et al., 1984; Carbery, 1984). A significant decline in serum copper with increasing age of mares appeared to be related to the incidence of usually fatal rupture of the uterine artery in aged, parturient mares (Stowe, 1968).

Horses are relatively tolerant of high dietary copper concentrations. Pony mares fed 791 mg of copper/kg of diet for 183 days had elevated liver copper, but no adverse clinical signs were observed in the mares or their foals (Smith et al., 1975). Single oral doses of 20 and 40 mg of copper/kg of body weight (as copper sulfate) were administered to mature ponies without apparent adverse effects (Stowe, 1980). The maximal tolerable level of copper for horses has been estimated to be approximately 800 mg/kg of diet (NRC, 1980).

FLUORINE

Fluorine is known to be involved in bone and teeth development in other species, but its dietary necessity for horses has not been established.

Signs of Excess Horses appear to be more tolerant than cattle of excess fluorine. Excess intake results in discolored teeth (fluorosis), bone lesions, lameness, and unthriftiness. Horses can tolerate 50 mg of fluorine/kg of diet for extended periods without detrimental effects (Shupe and Olson, 1971).

IODINE

Iodine is essential for the synthesis of thyroxine (T4) and triiodothyronine (T3), thyroid hormones that regulate basal metabolism. Iodine concentrations of some common equine feedstuffs range from 0 to 2 mg/kg, depending on the iodine content of the soil on which the feed was grown. Iodine supplementation can be accomplished by feeding iodized or trace mineralized salts that contain 70 mg of iodine/kg. The iodine requirement of horses was estimated to be 0.1 mg/kg of diet (Rodenwold and Simms, 1935). Normal values for T4 and T3 in horses are 5 to 30 and 0.45 to 1.5 ng/ml, respectively (R. Nachreiner, Michigan State University, personal communication, 1986).

Signs of Deficiency or Excess   Foals from iodine-deficient dams may be stillborn or born weak and have difficulty in standing to suckle even though the dams may not show signs of clinical deficiency. These foals commonly have a prominent enlargement of the neck over the anterior segment of the trachea due to thyroid hyperplasia (simple goiter). Iodine-deficient mares may exhibit abnormal estrous cycles (Kruzhova, 1968) but usually do not have thyroid enlargement.

Toxic dietary iodine concentrations may result from adding excessive supplemental iodine, such as ethylenediaminedihydroiodide (EDDI), to concentrates or from using feedstuffs high in iodine. A common feedstuff containing excess iodine is kelp, a seaweed that may contain as much as 1,850 mg of iodine/kg (Baker and Lindsey, 1968). Iatrogenic iodinism and an associated alopecia in a horse being treated with EDDI for dermatophilosis have been reported (Fadok and Wild, 1983).

The maximal tolerable dietary concentration of iodine has been estimated to be 5 mg/kg of dry matter (NRC, 1980), equivalent to 50 mg of iodine/day for a horse consuming 10 kg of dry matter daily. However, pregnant mares fed 35 to 48 mg of iodine/day gave birth to foals with enlarged thyroids due to iodine toxic goiter (Baker and Lindsey, 1968; Drew et al., 1975; Driscoll et al., 1978). Because the milk of these mares can be expected to contain excess iodine (Brown-Grant, 1957), an alternate source of milk low in iodine is necessary to permit the iodine-toxic foals to recover from the in utero effects of iodine excess. Thus, before iodine supplementation is provided to mares or their goitrous foals, it is important to establish whether the horses had been fed too little or too much iodine.

IRON

The body of a 500-kg horse contains about 33 g of iron. This is distributed in hemoglobin (60 percent), myoglobin (20 percent), storage and transport forms (20 percent), and cytochromic and other enzymes (0.2 percent) (Moore, 1951). Forage and by-product ingredients commonly contain 100 to 250 mg of iron/kg. Grains usually contain less than 100 mg/kg. Dietary iron absorption in nonruminants fed adequate iron is likely to be 15 percent or less. Iron utilization increases in iron-deficient diets and diminishes with higher than normal intakes of cadmium, cobalt, copper, manganese, and zinc (Underwood, 1977). Daily endogenous losses of iron have not been reported for horses, and their dietary iron requirement has not been researched specifically. The iron requirement is estimated to be 50 mg/kg of diet for growing foals or pregnant and lactating mares and 40 mg/kg of diet for mature horses. Common feedstuffs, therefore, should meet the iron requirements.

According to Meyer (1986), approximately 37, 38, and 92 mg of iron are deposited each day in the fetus and membranes during months 9, 10, and 11 of gestation, respectively. In a 500-kg mare, this translates to 74, 76, and 184 μg/kg of mare weight. The iron content of mare's milk ranges from 1.3 μg/g at parturition to 0.49 μg/g at 4 months postpartum (Ullrey et al., 1974). A mare producing 15 kg of milk/day would require approximately 130 mg of iron daily for early lactation and 32.6 mg of iron for 10 kg of milk/day in late lactation, in addition to the iron requirement for maintenance. Serum ferritin, an accurate measure of iron status, ranged from 70 to 250 ng/ml with a mean of 152 ± 54.6 for normal horses (Smith et al., 1984).

Signs of Deficiency or Excess   The primary signs of iron deficiency are microcytic and hypochromic anemia. Although young, milk-fed foals are most susceptible to this anemia, iron deficiency is not a practical problem in foals or mature horses at any performance level. This is true, in part, because the body efficiently salvages and retains iron derived from the catabolism of body constituents. Various iron supplements, under natural feeding programs, have been ineffective in improving the hemoglobin or oxygen-carrying capacity of red blood cells (Kirkham et al., 1971).

High concentrations of supplemental dietary iron (500 and 1,000 mg/kg) fed to ponies had no effect on feed intake, daily gain, red blood cell count, hemoglobin concentration, packed cell volume, or serum iron, calcium, copper, and manganese (Lawrence, 1986; Lawrence et al., 1987). The higher dietary iron level, however, depressed both serum and liver zinc.

Excess iron is especially toxic to young animals, and deaths among foals have been attributed to oral administration of digestive inocula containing supplemental iron (Mullaney and Brown, 1988). Foals dosed according to manufacturer's recommendations received 350 mg of elemental iron as ferrous fumarate at birth and at 3 days of age. Prior to death, these foals exhibited diarrhea, icterus, dehydration, and coma. Morphologic changes included erosion of jejunal villi, pulmonary hemorrhage, massive iron deposition in the liver, and liver degeneration. Ferrous fumarate toxicity in a mature horse has been reported by Arnbjerg (1981).

MANGANESE

Manganese is essential for carbohydrate and lipid metabolism and for synthesis of the chondroitin sulfate necessary in cartilage formation. The manganese requirements of horses have not been established; however, based upon data from other species, 40 mg of manganese/kg of diet is considered adequate (Rojas et al., 1965). Roughages contain 40 to 140 mg of manganese/kg, and concentrates (except corn) contain 15 to 45 mg/kg.

Signs of Deficiency or Excess Manganese deficiency in other species results in abnormal cartilage development. This is due to failure of chondroitin sulfate synthesis and results in bone malformation. The crooked limbs of newborn calves have been associated with manganese deficiency (Howes and Dyer, 1971). Similar afflictions (congenitally enlarged joints, twisted legs, and shortened forelimb bones) have been associated with "smelter smoke syndrome" in Oklahoma (Cowgill et al., 1980); the extensive liming required to offset the acidic effects of smelter effluent on the soil is thought to reduce the manganese availability markedly. It has been suggested since, but not proven, that manganese deficiency may be associated with limb abnormalities and congenital contractures in newborn foals. However, no direct evidence exists to support this theory.

SELENIUM

Selenium is an essential component of selenium-dependent glutathione peroxidase (Rotruck et al., 1973), which aids in detoxification of lipo- and hydrogen peroxides that are toxic to cell membranes.

The concentration of selenium in feedstuffs commonly ranges from 0.05 to 0.3 ppm and is influenced by variations in soil selenium and pH. Selenium is absorbed quite efficiently (77 percent) in nonruminants, in contrast to about 29 percent for ruminants (Wright and Bell, 1966). Selenium in forages and seed grains is normally present as organic selenium in the form of selenocystine, selenocysteine, and selenomethionine. Sodium selenite and sodium selenate are common inorganic sources of supplemental selenium. Although the Food and Drug Administration (FDA) has approved maximal selenium supplementation at 0.3 mg/kg of dry matter in complete feeds for cattle, sheep, and swine (FDA, 1987), selenium supplementation of equine feeds is restricted only by nutritional recommendations and industry practices.

The selenium requirement of the horse was estimated at 0.1 mg/kg of diet (Stowe, 1967). This is consistent with the report of Shellow et al. (1985), who found that plasma selenium concentrations of mature horses reached a plateau at about 140 ng/ml in horses fed either 0.14 or 0.23 mg of selenium/kg of diet. They concluded that there was no advantage in supplementing the mature idle horse with more than 0.1 mg of selenium/kg of diet and that 140 ng of selenium/ml of plasma (or serum) was adequate to prevent problems associated with selenium deficiency. A similar selenium supplementation rate, 1 mg/day for horses 1 to 6 years of age, was reported by Maylin et al. (1980) to increase blood selenium from 45 to 123 ng/ml over an 11-week period. The higher blood selenium value was considered well above the concentration associated with myodegeneration. Glutathione peroxidase (GSH-px) values for racing Standardbred horses were reported to be 17 enzyme units (EU)/mg of hemoglobin (Gallagher and Stowe, 1980). Maylin et al. (1980) and Roneus and Lindholm (1983) confirmed a strong relationship between selenium intake and GSH-px activity and noted that the GSH-px response to oral selenium was much lower than to parenteral selenium.

Signs of Deficiency or Excess The selenium status of horses can be evaluated by measuring serum, plasma, or whole blood selenium by the sensitive fluorometric selenium assay (Whetter and Ullrey, 1978). Selenium-dependent GSH-px of serum and erythrocytes can also be measured (Paglia and Valentine, 1967); however, sample storage time and temperature are critical. The clinical and morphologic manifestations of selenium deficiency are affected by the concomitant vitamin E status. Nutritional myopathy (white muscle disease or vitamin E/selenium-responsive disease) involves skeletal and cardiac muscles; it is associated with GSH-px values lower than 25 EU/dl (Caple et al., 1978) and with serum selenium values lower than 60 ng/ml (Blackmore et al., 1982). Acute, subacute, and chronic forms of selenium deficiency have been reported in China (Jiong et al., 1987). The myopathy results in weakness, impaired locomotion, difficulty in suckling and swallowing, respiratory distress, and impaired cardiac function (Dill and Rebhun, 1985). Serum changes include elevations in creatine kinase, aspartate aminotransferase, potassium, and blood urea nitrogen (Dill and Rebhun, 1985). Elevated aspartic-pyruvic transaminase and γ-glutamyltransferase have also been associated with the vitamin E/selenium-responsive disease. The tying-up syndrome did not correlate with vitamin E or selenium status (Lindholm and Asheim, 1971; Gallagher and Stowe, 1980; Blackmore et al., 1982).

The serum selenium of foals from selenium-adequate mares is typically much lower than their dams and ranges from 70 to 80 ng of selenium per ml of serum (Stowe, 1967). If, according to Blackmore et al. (1982), serum selenium values below 65 ng/ml are indicative of deficiency, young foals may be prone to nutritional muscular dystrophy, especially if their vitamin E status is low.

The maximal tolerable level of selenium in horses is estimated at 2 mg/kg of diet (NRC, 1980), and the LD_{50} for orally administered selenium is considered to be approximately 3.3 mg of selenium(as sodium selenite)/kg of body weight (Miller and Williams, 1940). Copper pretreatment can increase the LD_{50} markedly (Stowe, 1980). Acute selenium toxicity—blind staggers—is characterized by apparent blindness, head pressing, perspiration, abdominal pain, colic, diarrhea, increased heart and respiration rates, and lethargy (Rosenfeld and Beath, 1964). Chronic selenium toxicity—alkali disease—is characterized by alopecia, especially about the mane and tail, as well as cracking of the hooves around the coronary band (Rosenfeld and Beath, 1964; Traub-Dargatz and Hamar, 1986). There are anecdotal accounts of immediate death after administration of injectable vitamin E/selenium preparations. These deaths appear due to an anaphylactoid sensitivity of the horse to a carrier ingredient in the injectable preparations and not to the toxicity of selenium or vitamin E.

ZINC

Zinc is present in the body as a component of many metalloenzymes such as carbonic anhydrase, alkaline phosphatase, and carboxypeptidase. The biochemical role of zinc relates largely to the functions of these enzymes. The highest concentrations of zinc occur in the choroid and iris of the eye and in the prostate gland. Intermediate concentrations of zinc are present in the skin, liver, bone, and muscle, whereas low concentrations are found in blood, milk, lungs, and brain. Zinc absorption is regulated by the zinc status of the animal

and may be in the 5–10 percent range. Common equine feedstuffs contain 15 to 40 mg of zinc/kg. Appropriate sources of supplemental zinc include zinc sulfate, zinc oxide, zinc chloride, zinc carbonate, and various zinc chelates.

Harrington et al. (1973) demonstrated that 40 mg of zinc/kg of a purified-type diet was sufficient to prevent zinc deficiency in foals. Schryver et al. (1974) reported that foals fed 41 mg of zinc/kg of natural diet grew at acceptable rates and maintained normal body stores of zinc. Drepper et al. (1982) and Jarrige and Martin-Rosset (1981) indicated that 50 mg of zinc/kg of dry matter was adequate for all classes of horses. Mares' milk contains 1.8 to 3.2 mg of zinc/kg of fluid milk (Appendix Table 1). Thus, foals drinking 15 kg of milk/day would consume 27 to 48 mg of zinc/day. On a dry matter basis, this is equivalent to 17 to 30 mg of zinc/kg of dry matter intake. The zinc in milk is assumed to be highly available.

Signs of Deficiency or Excess Zinc deficiency has been produced in foals fed 5 mg of zinc/kg of purified diet (Harrington et al., 1973). Zinc deficiency in foals is accompanied by inappetence, reduced growth rate, parakeratosis (especially on the lower limbs), alopecia, reduced serum and tissue zinc concentrations, and decreased alkaline phosphatase (Harrington et al., 1973). According to Knight et al. (1985), based on farm surveys the optimal dietary zinc concentration in equine diets to minimize the incidence of metabolic bone disease approaches 90 mg/kg. Although these workers suggested that the previous NRC (1978) recommendations on zinc requirements be reevaluated, no controlled studies support a dietary zinc requirement greater than 50 mg/kg of diet DM.

Horses appear quite tolerant of excess dietary zinc. No detrimental effects were observed in mares or foals fed diets containing up to 700 mg of zinc/kg (Graham et al., 1940). However, Messer (1981) reported tibiotarsal effusion in three Arabian fillies that had marked elevations in serum zinc. Foals fed 90 g of zinc/day (equal to about 2 percent of the diet) developed enlarged epiphyses, stiffness of gait, lameness, and increased tissue zinc (Graham et al., 1940). Similar signs were observed by Eamens et al. (1984) in four young horses grazed near industrial plants where the herbage used as pasture contained high zinc concentrations and by Gunson et al. (1982) in two foals raised near a zinc smelter. The cause appeared to be a secondary copper deficiency induced by zinc toxicosis. Young et al. (1987) and Coger et al. (1987) were unable to significantly alter the copper absorption in growing and mature ponies fed diets containing 580 or 1,200 mg of zinc/kg. Therefore, the purported effect of elevated zinc intake on copper metabolism may involve postabsorptive events rather than the actual site of absorption.

VITAMINS

Vitamin requirements, like those of other nutrients, are affected by age, stage of production, and a variety of stresses such as gastrointestinal infections and intense muscular exercise. The need for supplemental vitamins depends on the type and quality of the diet, the amount of microbial vitamin synthesis in the digestive tract, and the extent of vitamin absorption from the site of synthesis. Horses grazing high-quality pastures are likely to need little or no vitamin supplementation because forages are a rich source of most fat- and water-soluble vitamins.

Fat-Soluble Vitamins

The fat-soluble vitamins include vitamin A and carotenes, vitamin D, vitamin E, and vitamin K. Diseases or metabolic abnormalities that interfere with fat absorption also adversely affect absorption of these vitamins. Oral administration of fats that are not absorbed, such as mineral oil, may result in undesirable loss of fat-soluble vitamins in the feces. Anaerobic bacteria in the intestinal tract can synthesize vitamin K, but other fat-soluble vitamins must be present in the diet or, for vitamin D, the horse must be exposed to ultraviolet light. Quantitative requirements are usually expressed in International Units (IU) or United States Pharmacopeia (USP) units for vitamins A, D, and E and in milligrams for vitamin K.

VITAMIN A

The term vitamin A is a generic descriptor for all derivatives of β-ionone (other than the carotenoids) that possess the biological activity of all-*trans*-retinol. One IU or USP unit of vitamin A activity can be derived from 0.3 μg of all-*trans*-retinol. At least 12 major forms or derivatives of retinol occur in nature. The major commercial forms of vitamin A are all-*trans*-retinyl palmitate (1 IU = 0.549 μg) and all-*trans*-retinyl acetate (1 IU = 0.344 μg). They are more stable than all-*trans*-retinol, are more soluble in oils, and can be embedded in a gelatin matrix with antioxidants to protect them from destruction over an extended period.

Vitamin A is important for vision, and specific photoreactive metabolites are found in visual pigments within the retina. This vitamin also plays a basic role in cell differentiation, and in a vitamin A deficiency, epithelial cells that would normally be of a transitional, cuboidal, or columnar type may develop as keratinized squamous cells. Bone remodeling in the growing animal is also modulated by vitamin A.

The term provitamin A is a generic descriptor for all carotenoids that show the biological activity of vitamin A. Over 400 carotenoids have been isolated from nature, but only about 50 have vitamin A activity. The most active and quantitatively important is all-*trans*-β-carotene. One IU of provitamin A activity can be derived from 0.6 μg of all-*trans*-β-carotene by the rat or chick under very specific conditions. The horse, however, is not as efficient. All-*trans*-β-carotene is found in highest concentration in green forage but is also present in yellow corn. Because carotenes are destroyed gradually by light and heat, the carotene concentration (on a dry basis) in sun-cured hay is lower than in the fresh forage from which the hay was made. If the hay is stored, the carotene concentration declines gradually until, after 2 years, less than 10 percent of the original carotene level may remain (Waite and Sastry, 1949). In addition, if forage is rained on and the period of field curing is extended, appreciable leaf loss may occur with a serious decrease in carotene even before storage. Meigs (1939) reported that carotenes in fresh-growing pasture grass and alfalfa may reach 300 to 600 mg/kg of dry matter, whereas carotenes in U.S. No. 1 alfalfa hay and U.S. No. 1 timothy hay constitute about 40 and 20 mg/kg of dry matter, respectively. Carotenes in U.S. No. 3 alfalfa hay and U.S. No. 3 timothy hay are only about 4 to 5 mg/kg of dry matter.

Garton et al. (1964) noted that mean plasma concentrations of carotene and vitamin A in 24 adult mares were 7.9 and 15.8 μg/dl, respectively, in winter on hay, and 114.1 and 26.7 μg/dl, respectively, on early spring pasture. By the middle of summer, plasma carotene and vitamin A concentrations had declined somewhat, to be followed by a slight increase in the fall. During the succeeding winter, respective concentrations decreased once again to 15.6 and 14.5 μg/dl of plasma. Although a mean carotene concentration was reported only for summer and fall pasture (204 mg/kg of dry matter), seasonal differences in plasma carotene vitamin A concentrations were probably associated with differences in forage carotene concentration.

Mäenpää et al. (1988a) reported a similar seasonal pattern in vitamin A concentration in serum collected monthly from mares and foals of the Finnhorse breed. Mares and foals were kept on pasture from early June until early October, when they were stabled and fed timothy hay and oats until the end of May. Serum retinol concentrations in mares on pasture rose rapidly from a low in May of about 15 μg/dl to highs in June through October of about 23 to 28 μg/dl. From November to May, serum retinol concentrations in mares declined gradually. Foals had lower serum retinol concentrations than mares, with values of about 13 μg/dl near birth in June, rising to 18 μg/dl in September, declining to a low of 11 μg/dl in April, and rising again to 23 μg/dl in June when they became yearlings. When mares were given daily supplements of 10,000 to 40,000 IU of vitamin A, serum retinol concentrations did not increase (Mäenpää et al., 1988b). However, when weanling foals were supplemented with 40,000 IU of vitamin A/day, serum retinol concentrations in late winter and early spring increased, but not to levels found in summer.

Much of the conversion of dietary carotene to vitamin A presumably occurs in the wall of the small intestine through the action of β-carotene 15,15′-dioxygenase. However, this conversion appears not to be very efficient in horses, and a difference in the utilization of carotenes from various forage species may occur. Fonnesbeck and Symons (1967) found that hay supplying carotene in average amounts of 198 mg/day to Standardbred geldings weighing 384 to 474 kg did not maintain initial plasma vitamin A concentrations. By the end of a 24-week study begun in January in New Jersey, mean plasma vitamin A values had declined from 14.6 to 8.5 μg/dl. During this period, plasma carotene concentrations declined from 8.5 to 1.9 μg/dl. Bromegrass, fescue, canarygrass, red clover, and alfalfa hays were among the forages tested. Carotene in alfalfa hay appeared to be more available as a source of provitamin A activity than carotene in grass hays. Carotene intake from alfalfa hay was approximately 24.5 mg/kg of dietary dry matter or 400 to 500 μg/kg of body weight.

These values contrast with the findings of Guilbert et al. (1940), who worked with young Percherons and concluded that the minimum dietary carotene requirements for horses were 20 to 30 μg/kg of body weight. They used dehydrated alfalfa meal as the source of carotene, and the minimal requirement was defined as the lowest level of carotene that prevented nyctalopia. They concluded also that 20 to 30 μg of carotene/kg of body weight supported normal growth and freedom from clinical signs of deficiency, but supported little or no carotene or vitamin A storage. Minimal carotene requirements for significant tissue storage, optimal dark adaptation, and reproduction were proposed to be five times greater, or approximately 125 μg/kg of body weight. However, this conclusion appeared to be based on research with other species, and data on horses were not provided. In addition, the various carotenoid isomers undoubtedly present in the feedstuffs studied were not quantified. Thus, differences in the biological activity of these isomers could account for some of the discrepancy between the studies.

When Guilbert et al. (1940) estimated equine dietary vitamin A requirements using cod liver oil as the vitamin A source, they concluded that the minimal requirement to support normal growth, freedom from clinical signs, and little or no tissue storage was 20 IU/kg of body weight. For significant tissue storage, optimal dark adaptation, and reproduction, a requirement of 60 IU/kg of body weight was proposed.

Other workers have reported estimates of carotene or vitamin A need. Ralston et al. (1985) concluded that 17 to 19 mg of β-carotene/kg of dry matter in grass hay was adequate for semen production and libido in stallions, and Stowe (1968b) noted that 9.5 to 11 IU of vitamin A/kg of body weight prevented signs of deficiency. Based on interpolations of the response of cellular hematology, serum biochemistry, and tissue vitamin A in ponies to dietary vitamin A providing 0, 12, 1,200, or 12,000 μg of retinol/kg of body weight/day, Donoghue et al. (1981) proposed that an optimal range might be 18 to 60 μg (60 to 200 IU) of retinol/kg of body weight/day.

A number of studies dealing with dietary carotene and female reproduction have been published. Ahlswede and Konermann (1980) reported that horses on pasture had plasma β-carotene concentrations 8 to 13 times higher than horses kept in stables. They also reported that carotene supplementation of stabled horses tended to improve ovarian activity. Van der Holst (1984) reported that β-carotene supplementation produced stronger heats, improved conception, and tended to reduce embryonic mortality in pony mares. Ferraro and Cote (1984) suggested that feeding 100 mg of β-carotene/day to horses seemed to induce earlier and stronger heats, improve conception rates, and aid in maintenance of pregnancy. However, Eitzer and Rapp (1985) found no benefit from β-carotene supplements of 400 mg/day to mares fed a diet providing 70 to 80 mg of β-carotene/day. The mares fed the control diet had plasma β-carotene levels (about 10 μg/dl) similar to those observed by Ahlswede and Konermann (1980) in stabled horses. Although this issue is not yet settled, β-carotene supplements appear unlikely to be beneficial if mares are on pasture or are fed forages containing high levels of carotene.

Because of the limited data available on carotene or vitamin A requirements and their variability, only imprecise estimates of need can be made. Diets for all horses should provide 30 to 60 IU (9 to 18 μg) of vitamin A activity as retinol or equivalent/kg of body weight/day. If vitamin A activity is supplied by dietary β-carotene, 120 to 240 IU of provitamin A activity (72 to 144 μg of β-carotene)/kg of body weight/day should be provided. Thus, 1 mg of β-carotene ought to be considered equivalent to no more than 400 IU of vitamin A activity.

Signs of Deficiency or Excess   A deficiency of vitamin A is characterized by night blindness, excessive lacrima-

tion, hyperkeratinization of the cornea and skin, anorexia, poor growth, respiratory infections, sublingual
salivary gland abscesses, impaired conception, elevated
cerebrospinal fluid pressure, convulsive seizures, pro-
gressive weakness, and declines in plasma, liver, and
kidney vitamin A concentrations (Edwards, 1937; Howell et al., 1941; Stowe, 1968b; Donoghue et al.,
1981).

Prolonged feeding of excess vitamin A may cause bone
fragility, hyperostosis, exfoliated epithelium, and tera-
togenesis. Severe intoxication of ponies (12,000 μg of re-
tinol/kg of body weight/day) produced unthriftiness by
15 weeks and, shortly thereafter, rough hair coats, poor
muscle tone, and depression (Donoghue et al., 1981). By
week 20, ponies had lost large areas of hair and epidermis, were periodically ataxic, were severely depressed,
and spent much time in lateral recumbency. Plasma and
liver vitamin A concentrations were elevated, and
erythrocyte counts and packed red cell volumes were depressed. Jarrett et al. (1987) reported plasma retinol and
retinyl acetate concentrations in horses fed excess vitamin A in the form of retinyl palmitate. Compared to
horses fed 10,000 IU of retinyl palmitate/day for 30
days, horses fed 80,000 IU of retinyl palmitate exhibited
a shift in the proportions of serum retinol, retinyl acetate, and retinyl palmitate from 66, 4, and 30 percent to
43, 4, and 57 percent, respectively. However, the serum
concentration of total vitamin A-active compounds did
not change.

The NRC (1987) has proposed a presumed upper safe
level of 16,000 IU of vitamin A/kg of dry diet for chronic
administration. If a 500-kg horse consumes dietary dry
matter at 2 percent of its body weight, this is equivalent
to 320 μg (1,067 IU) of all-*trans*-retinol/kg of body
weight/day. If one calculated provitamin A intake from
10 kg/day of fresh alfalfa dry matter when 600 mg of
carotene/kg of dry matter is present, 12,000 μg of caro-
tene/kg of body weight would be supplied. Although
this number is large, no carotene toxicity has ever been
demonstrated on alfalfa pasture, and the efficiency of
carotene conversion to vitamin A under these circumstances is probably reduced due to regulation of β-carotene 15,15'-dioxygenase activity.

VITAMIN D

Two primary vitamin D-active compounds are found
in nature. Both are a result of photobiogenesis from exposure to solar wavelengths of 290 to 315 nm. Vitamin
D_2 (ergocalciferol) results from ultraviolet irradiation of
ergosterol, which is synthesized by plants. This irradiation is not effective, however, in living plant tissue, presumably because chlorophyll screens out the necessary
wavelengths. Thus, vitamin D_2 is found in plants after

they have been cut and exposed to sunlight, and in the
lower, dead leaves of living, insolated plants. Small
amounts of vitamin D_3 have also been identified in irra-
diated alfalfa (Horst et al., 1984). Based on studies of
other species, it has been presumed that vitamin D metabolism in the horse is as follows. Vitamin D_3 (cholecal-
ciferol) results primarily from ultraviolet irradiation of
7-dehydrocholesterol synthesized by the tissues of the
horse and present in the skin. When absorbed from the
intestinal tract (vitamin D_2 or D_3) or translocated from
the skin (vitamin D_3), both vitamins are hydroxylated
on the carbon in position 25 (C-25) in the liver and on
C-1 in the kidney. The hormonally active metabolites,
1,25-dihydroxyergocalciferol and 1,25-dihydroxychole-
calciferol, act on intestine, bone, and kidney to maintain calcium homeostasis. One mechanism by which
these metabolites act is to promote synthesis of calcium-
binding proteins. Calcium absorption from the intestine, resorption of calcium from the skeleton, and reabsorption of calcium in the distal renal tubule are
promoted. However, serum concentrations of 25-hy-
droxy-D_3 and 1,25-dihydroxy-D_3 are low in the horse
compared to some other species (Mäenpää et al., 1988a,
b). Serum 1,25-dihydroxy-vitamin D_3 concentrations
are similar in winter and summer (19 pg/ml), whereas
concentrations of 25-hydroxy-D_3 in the serum are higher
in summer (6 ng/ml) than in winter (4 ng/ml).

Dietary vitamin D requirements for the horse have
not been established, and deficiencies on a practical diet
are not likely. Young Shetland ponies kept outdoors in
Florida clearly do not need dietary vitamin D (El
Shorafa et al., 1979). However, when they were deprived of sunlight and no vitamin D was included in the
diet, ponies lost their appetite and had difficulty in
standing. Oral supplements of 1,000 IU of vitamin
D/day (form not indicated) prevented these signs, although ash (as a percentage of dry, fat-free bone), bone
density, breaking strength, and cross-sectional cortical
area tended to be lower than in ponies exposed to sunlight. Daily supplements of 1,000 IU of vitamin D were
equivalent to 781 IU/kg of diet for ponies 3 to 8 months
of age and 433 IU/kg of diet for ponies 9 to 14 months
old. Based on average body weights during the 5-month
study, these amounts of vitamin D translate to 28 and 9
IU/kg of body weight, respectively. Although limited,
these data suggest that for early growth, ponies deprived
of sunlight may require 800 to 1000 IU of vitamin D/kg
of dry diet. For later growth, 500 IU of vitamin D/kg of
dry diet may be sufficient.

Signs of Deficiency or Excess Dietary vitamin D supplements have been shown to promote calcium and
phosphorus absorption in horses (Hintz et al., 1973).
Nevertheless, Park (1923) stated that rickets occurs in-
frequently in horses, and Nieberle and Chors (1954)

stated that rickets never occurs in this species. Although El Shorafa et al. (1979) did not observe classical rickets in young ponies deprived of sunlight and dietary vitamin D (perhaps because the ponies stood very little), they found a loss of appetite, and slower growth, as well as decreased ash, cross-sectional cortical area, density, and breaking strength of metacarpal bones.

Excessive intake of vitamin D is characterized by calcification of blood vessels, heart, and other soft tissues and by bone abnormalities (Bille, 1970). Hintz et al. (1973) fed 3,300 IU of vitamin D_3/kg of body weight/ day to young ponies and produced death within 4 months. At necropsy, soft tissue calcification, generalized bone resorption, and extensive kidney damage were observed. Harrington (1982) gave 21 daily oral doses of either 9,300, 22,200, or 47,200 IU of vitamin D_2/kg of body weight to one 18-month-old horse per treatment. The horses on the two lower dosages did not exhibit clinical abnormalities during the study, but by the fifth day, the horse receiving 47,200 IU/kg of body weight/day appeared depressed, developed limb stiffness, and was reluctant to move. Interest in feed began to decline, and the horse was completely anorectic by day 16. It became recumbent on day 19, and 15 percent of its initial weight was lost by day 21. Marked hyperphosphatemia (7 to 13 mg of inorganic phosphorus/dl of serum) occurred in all horses. Intermittent hypercalcemia (> 13.3 mg of calcium/dl of serum) resulted from the low and high dosages but did not persist. All horses were killed between days 22 and 24. Mineralization of soft tissues occurred, particularly in the endocardium and the wall of large blood vessels, in horses receiving the high dose.

Because no clinical signs or lesions were observed in the horse receiving daily doses of 22,200 IU of vitamin D_2/kg of body weight in the study discussed above, Harrington and Page (1983) were surprised to find evidence of toxicity in two horses accidentally fed a diet providing 12,000–13,000 IU of vitamin D_3/kg of body weight/day. Thus, a subsequent study was conducted in which 30 daily oral doses of 33,000 IU of vitamin D_2/kg of body weight were fed to one horse and similar doses of vitamin D_3 were fed to a second. Signs of toxicity were evident in each case, including weight loss, hypercalcemia, hyperphosphatemia, and cardiovascular calcinosis. However, the signs of illness, deviations of blood chemistry from normal, and severity of tissue pathology were much more pronounced in the horse receiving vitamin D_3.

In addition to inadvertent additions of excess vitamin D to horse diets, naturally occurring toxic levels of vitamin D-like compounds have been found in certain plants. Krook et al. (1975) described hypercalcemia and calcinosis in horses consuming day jasmine (Cestrum diurnum) while grazing Florida pastures. The disease was clinically and morphologically similar to poisoning by a member of the nightshade family, duraznillo (Solanum glaucophyllum, also known as S. malacoxylon), seen in Argentina and Brazil (Worker and Carrillo, 1967; Dobereiner et al., 1971). The active calcitropic principle in both plants appears to be 1,25-dihydroxy-D_3-glycoside (Wasserman et al., 1976; Hughes et al., 1977), which results in hyperabsorption of calcium by animals eating these plants.

A maximum safe level of 2,200 IU of vitamin D_3/kg of diet has been proposed (NRC, 1987) for long-term feeding (greater than 60 days). For a 500-kg horse consuming dietary dry matter at 2 percent of body weight/day, this would be equivalent to 44 IU/kg of body weight/ day.

VITAMIN E

The term vitamin E is a generic descriptor for the natural or synthetic compounds that are qualitatively equivalent in biological activity to α-tocopherol. Eight or more such compounds are found in nature, including α-, β-, γ-, and δ-tocopherol and α-, β-, γ-, and δ-tocotrienol. Although not established for the horse, α-tocopherol seems to be the most biologically active, based largely on prevention of resorption of fetuses in the rat.

Biological activity is commonly expressed in IU even though no international standard exists. The USP standard, d,l(all-rac)-α-tocopheryl acetate, is approximately equivalent to the former IU; 1 mg of this compound has 1 USP unit (or 1 IU) of vitamin E activity. Commonly accepted equivalencies (USP units or IU/ mg) for other compounds are as follows: d,l(all-rac)-α-tocopherol, 1.10; $d(R,R,R)$-α-tocopheryl acetate, 1.36; $d(R,R,R)$-α-tocopherol, 1.49. The natural forms of α-tocopherol and its esters have the d (or R,R,R) configuration. Synthetic forms usually have the d,l (or all-rac) configuration.

Observations on myopathies that responded to vitamin E or selenium suggest an interrelationship between these two nutrients. Studies of this interrelationship led to the current understanding that vitamin E and selenium (in glutathione peroxidase) function as a part of a multicomponent antioxidant defense system. Vitamin E is located primarily in lipophilic parts of the cell, such as membranes, whereas glutathione peroxidase is found largely in the cytosol and mitochondrial matrix. These and other components of the defense system protect cells against the adverse effects of reactive oxygen forms and other free radical initiators of the oxidation of unsaturated phospholipids and of certain critical proteins. To function effectively as antioxidants, vitamin E compounds must be easily oxidized; as a consequence, naturally occurring forms are rather unstable. Commercial forms that have been esterified chemically, such as $d(R,R,R)$- or $d,1$(all-*rac*)-α-tocopheryl acetate, do not function as antioxidants (or oxidize) until the ester linkage has been split by esterases in the digestive tract. As a consequence, tocopheryl esters are quite stable.

Thus, the vitamin E activity of feeds and feedstuffs depends on the chemical forms present and the conditions of storage. Moisture in feeds sufficient to permit fermentation and mold growth results in rapid decline in the activity of natural vitamin E compounds. So do the grinding and storage of ground grains as opposed to storage as whole kernels. Grinding disrupts the internal architecture of the seed and exposes unsaturated lipids to air, thus promoting peroxidation.

Although vitamin E is required in the equine diet, little information is available from which to estimate quantitative need. Foals deficient in vitamin E required 27 μg of parenteral (intramuscular) or 233 μg of oral α-tocopherol/kg of body weight/day to maintain erythrocyte stability (Stowe, 1968a). These estimates were based on the period of time during which a single intramuscular or oral dose of α-tocopherol prevented layering hemolysis of erythrocytes from previously deficient animals. From these data, 1 mg of α-tocopherol administered intramuscularly appeared to be equivalent to 8.6 mg administered orally.

Roneus et al. (1986) estimated vitamin E requirements of adult, idle, Standardbred horses by tissue depletion-repletion techniques. All horses were depleted for 2 months on a diet supplying 107 mg of d,l-α-tocopheryl acetate equivalents daily. The horses were then divided into four groups of three horses each and fed the basal diet supplemented with 0, 200, 600, or 1,800 mg of d,l-α-tocopheryl acetate/day. This regimen continued for 112 days, except for the 200-mg treatment which was increased to 5,400 mg/day after 2 months when it became obvious that only small, insignificant changes in the vitamin E status of blood, liver, and skeletal muscle had occurred. This group received the increased supplement for an additional 106 days without any preliminary depletion. After the periods of supplementation, there was a second depletion period. Vitamin E concentrations (primarily α-tocopherol) in serum, liver, skeletal muscle, and adipose tissue at different stages of repletion-depletion were used to estimate vitamin E requirements. These workers concluded that to ensure nutritional adequacy, adult, idle, Standardbred horses fed a low-vitamin E diet should receive a daily, oral supplement of 600 to 1,800 mg of d,l-α-tocopheryl acetate. This would be equal to 1.5 to 4.4 mg/kg of body weight. Because serum, liver, and muscle vitamin E concentrations declined so rapidly during postsupplemental depletion, it was suggested that vitamin E supplements be provided daily rather than at less frequent intervals.

Mäenpää et al. (1988a) observed seasonal differences in serum α-tocopherol concentrations in mares and foals kept on pasture from early June until early October, then fed timothy hay and oats during the winter. Serum α-tocopherol concentrations were highest in August and September (about 2.7 μg/ml for mares, and 2.1 μg/ml for foals) and reached their lowest values in April or May (about 1.5 μg/ml for mares, and 1.2 μg/ml for foals). When mares were given winter supplements of 100 to 400 mg of vitamin E/day, no significant increases in serum α-tocopherol concentrations occurred and seasonal differences persisted. However, when foals were given supplements of 400 mg of vitamin E/day in the winter, serum α-tocopherol concentrations increased significantly during late winter and early spring to approximately 2 μg/ml in April and May from low values of about 1.3 μg/ml in December.

The effect of exercise on equine vitamin E requirements has not been established specifically, but Jackson et al. (1983) have shown that skeletal muscles from vitamin E-deficient rats and mice are more susceptible than normal muscles to contractile damage. Exercise endurance is determined largely by the functional mitochondrial content of muscle, and in rats, exercise to exhaustion results in decreased mitochondrial respiratory control, loss of sarcoplasmic and endoplasmic reticulum integrity, and increased concentrations of lipid peroxidation products and free radicals in both liver and muscle (Davies et al., 1982). It is relevant that the exercise endurance of vitamin E-deficient rats was 40 percent lower than that of controls (Quintanilha and Packer, 1983). Janssen and Ullrey (unpublished research, 1986) observed that the capture and restraint of zebra and Przewalski horses for hoof trimming resulted in temporary to persistent muscle soreness and lameness when diets provided approximately 50 IU of vitamin E/kg. Plasma α-tocopherol concentrations were commonly below 1 μg/ml. When dietary vitamin E concentration was increased to approximately 100 IU/kg of dry matter, no clinical signs of muscle pathology were seen, and plasma α-tocopherol concentrations increased to 1.5 to 3 μg/ml. Butler and Blackmore (1983) reported that the mean (± SD, the standard deviation) plasma or serum α-tocopherol concentration for 140 samples from stabled Thoroughbreds in training was 3.3 ± 1.29 μg/ml. A dietary concentration of 100 IU of vitamin E/kg of dry matter provides the equivalent of about 2 mg of d,l-α-tocopheryl acetate/kg of body weight/day. Because the above information is limited and different criteria of adequacy support different conclusions, establishing a single minimum requirement for vitamin E is difficult. One criterion that has not been examined in the horse in sufficient detail is the amount of vitamin E required for optimum immune function. Baalsrud and Overnes (1986) supplemented a basal diet daily with 600 IU of vitamin E, 5 mg of selenium, or both. They observed an improved humoral immune response to tetanus toxoid or equine influenza virus in adult horses supplemented with selenium alone or selenium plus vitamin E. No improvement was seen with vitamin E alone, although an insufficient amount of supplement may have been used. Bendich et al. (1986) concluded that the dietary vitamin E requirement for optimum immune response in the rat exceeds 50 mg (all-*rac*)-α-tocopheryl acetate/kg of air-dry diet and may be as high as 100 mg/kg or more. If this is true for the horse, vitamin E requirements may exceed 50 IU/kg of dry diet or 1 IU/kg of body weight. Obviously, more research on this subject is needed. Until this issue is clarified, it may be prudent to ensure that diets contain 80 to 100 IU of vitamin E/kg of dietary dry matter in the total diet for foals, pregnant and lactating mares, and working horses.

Signs of Deficiency or Excess   Although it is difficult to distinguish between the morphological lesions of vitamin E and selenium deficiencies, signs representing a possible deficiency of both nutrients have been described in the foal (Schougaard et al., 1972; Wilson et al., 1976). Myodegeneration was common, with pale diffuse or linear areas in skeletal and cardiac muscle. Histological examination revealed hyaline and granular degeneration, as well as swelling and fragmentation of muscle fibers from several sites, including the tongue. The latter defect may interfere with normal nursing. Subcutaneous and intramuscular edema, pulmonary congestion, and occasionally, steatitis were also observed.

Liu et al. (1983) reported a degenerative myelopathy in six Przewalski horses from 13 months to 14 years of age. They had been fed commercially prepared horse pellets, timothy hay, and fresh grass in the summer. The horse pellets contained 22 IU of vitamin E and 0.3 mg of selenium/kg. Plasma α-tocopherol concentrations were low and ranged from less than 0.3 to 0.8 μg/ml. Hepatic selenium concentration in the one horse whose value was determined was 1.47 μg/g of dry matter. Ataxia was evident in all, including uncoordinated movement of the hind limbs, and an abnormally wide-based gait and stance. No gross lesions were observed in the brain, vertebrae, or spinal cord, but histologic examination revealed degeneration of the neural processes in the ventral and lateral funiculi. Myelin sheaths were dilated and vacuolated, and a number of axons were swollen, fragmented, or lysed. Neuronal degeneration, phagocytosis, and accumulation of periodic acid-Schiff-positive, xylol-insoluble lipopigment occurred in the affected neurons of the dorsal root ganglia.

Mayhew et al. (1987) described a vitamin E-responsive degenerative myeloencephalopathy in Standardbred and Paso Fino horses, from 3 to 30 months old. Symmetric ataxia and paresis, along with laryngeal adductor, cervicofacial, local cervical, and cutaneous trunci hyporeflexia, were characteristic. No clinical signs were observed before 3 months of age. The onset of gait abnormality was usually abrupt. Clinical signs then remained static or progressed for weeks or months. Severely affected animals often fell while running. The mean (± SD) serum α-tocopherol concentration of 13 ataxic weanlings was 0.62 ± 0.13 (range 0.47 to 0.84) μg/ml. A number of nonataxic weanlings had similar values, although vitamin E supplementation markedly reduced the incidence of the syndrome on affected farms. Based on genetic studies, this disorder appeared to have a familial disposition.

Signs of vitamin E toxicity in the horse have not been produced. Dietary levels of at least 1,000 IU/kg can apparently be fed for prolonged periods without harm to rats or chicks (NRC, 1987). Based on studies by March et al. (1973), the presumed upper safe level for chicks is between 1,000 and 2,000 IU/kg. Studies by Yang and Desai (1977a,b) and Alam and Alam (1981) indicated that rats can tolerate 2,500 IU/kg of diet. The NRC (1987) has proposed a safe upper level for animals of about 75 IU/kg of body weight/day. However, this would provide 37,500 IU/day for a 500-kg horse. If this horse consumed dry matter at 2 percent of its body weight/day, the vitamin E concentration in the diet dry matter would be 3,750 IU/kg. Because very high intakes of vitamin E may interfere with utilization of other fat-soluble vitamins and even induce a coagulopathy in the mildly vitamin K-deficient dog (Corrigan, 1979), a more conservative maximum tolerable level would be 1,000 IU/kg of dry diet or 20 IU/kg of body weight/day.

VITAMIN K

The term vitamin K is a generic descriptor for 2-methyl-1,4-naphthoquinone (menadione, vitamin K_3) and 3-substituted derivatives that exhibit antihemorrhagic activity. The compound found in green plants, 2-methyl-3-phytyl-1,4-naphthoquinone, is generally called vitamin K_1 or phylloquinone. Vitamin K compounds synthesized by bacteria (once thought to be a single compound known as vitamin K_2) constitute a series with unsaturated polyisoprenoid side chains at the 3-position and are known as menaquinones. Predominant bacterial forms contain a side chain of 6 to 10 isoprenoid units, but may contain up to 13, and several forms have partially saturated side chains.

The function of this vitamin is expressed through the activity of a vitamin K-dependent carboxylase found in the microsomal fraction of horse liver, spleen, and kidney (Vermeer and Ulrich, 1982), which promotes the posttranslational conversion of specific glutamyl residues in vitamin K-dependent plasma proteins to γ-carboxyglutamyl residues. These residues are essential for the Ca^{++}-dependent interaction of vitamin K-dependent clotting factors with phospholipid surfaces.

Dietary vitamin K requirements have not been determined for the horse. Phylloquinone in pasture or in good-quality hay and menaquinones synthesized by intestinal bacteria presumably meet those requirements in all but the most unusual of circumstances.

Signs of Deficiency or Excess If the intake or intestinal synthesis of vitamin K is insufficient, the rate of thrombin production decreases. This, in turn, results in decreased fibrin clot formation and increased susceptibility to hemorrhage.

Excess intake of phylloquinone appears to be essentially innocuous. Molitor and Robinson (1940) administered 25 g/kg of body weight orally or parenterally to laboratory animals with no adverse effect. Menaquinones and menadione in the diet probably also have low toxicity. The NRC (1987) proposed that oral toxic levels are at least 1,000 times the dietary requirement. However, Rebhun et al. (1984) administered single doses of menadione bisulfite to horses in amounts of 2.1 to 8.3 mg/kg of body weight via intramuscular or intravenous routes. These dosages conformed to manufacturers' recommendations but resulted in renal colic, hematuria, azotemia, and electrolyte abnormalities consistent with acute renal failure. At necropsy, lesions of renal tubular nephrosis were found. Because phylloquinone injectables appear safer than menadione injectables for the human newborn (American Academy of Pediatrics, 1971), use of the former seems preferable when parenteral vitamin K is administered to the horse.

Water-Soluble Vitamins

The B-complex vitamins, with the exception of vitamin B_{12}, are usually supplied in adequate amounts in good-quality forage. Because vitamin B_{12} is not synthesized by higher plants but is synthesized by anaerobic bacteria (provided sufficient cobalt is present), microbial synthesis in the cecum and colon can meet tissue needs. Abundant synthesis of other B vitamins by the intestinal microflora also occurs, and this source, plus the food supply, will meet the B vitamin requirements of most adult horses. The situation with young horses has been little explored. It is assumed that mare's milk generally meets the B vitamin needs of the nursing foal, and other feeds contribute as well once the foal begins to eat forage and grain. If artificial diets must be formulated for the orphan foal, the following NRC (1988) requirements per kg of diet for the baby pig can be considered (on a 100 percent dry matter basis): biotin, 0.09 mg; choline, 0.7 g; folacin, 0.3 mg; niacin (available), 22 mg; pantothenic acid, 13 mg; riboflavin, 4 mg; thiamin, 2.0 mg; vitamin B_6, 2.0 mg; and vitamin B_{12}, 22 μg.

THIAMIN

Thiamin synthesis in the intestinal tract of horses was inferred by Carroll et al. (1949) who found the following concentrations of thiamin (mg/kg of dry matter) in the diet and in intestinal contents: diet, 1.1; duodenum, 0.5; ileum, 2.2; cecum, 7.1; anterior large colon, 17.8; and anterior small colon, 7.8. Linerode (1967) estimated that 25 percent of the free thiamin in the cecum is absorbed by the horse. However, absorption of microbially synthesized thiamin in the intestine may not meet total needs, and several researchers have demonstrated a dietary requirement (Naito et al., 1925; Carlstrom and Hjarre, 1939; Pearson et al., 1944a; Carroll et al., 1949).

Carlstrom and Hjarre (1939) found that a daily thiamin intake of 3 μg/kg of body weight was inadequate, whereas 55 μg/kg of body weight maintained normal appetite, weight gain, and thiamin concentration in skeletal muscle. If the intake of dietary dry matter is 2 percent of body weight/day, these amounts of thiamin would be comparable to 0.15 and 2.75 mg/kg of dietary dry matter, respectively. Carroll (1950) made a similar observation and found that 3 mg of thiamin/kg of air-dry diet was sufficient. Jordan (1979) fed a pelleted concentrate-roughage mixture to twelve 3- to 4-month-old ponies divided equally between a basal diet and one supplemented with 6.6 mg of thiamin/kg of air-dry matter. Although the basal diet was not assayed for thiamin, calculations from the NRC (1982) feed composition tables yield an estimate of 2.6 mg of thiamin/kg of air-dry diet or 2.9 mg/kg of diet dry matter. Weight gains over 132 days were significantly greater among the ponies fed the thiamin supplement compared to those fed the unsupplemented basal diet. It should be noted that gastrointestinal transit time of a pelleted concentrate-roughage mixture is shorter than if the items were not ground. Thus the opportunity for microbial thiamin synthesis in the gut decreases.

Effects of thiamin supplements on exercising horses have been explored by Topliff et al. (1981) who proposed, based on measures of blood lactate, pyruvate, thiamin, and thiamin balance, that 4 mg of thiamin/kg of air-dry diet may not be sufficient.

The presence of thiaminases and antithiamin compounds in dietary plants such as bracken fern (*Pteridium aquilinum*), horsetail (*Equisetum arvense*), and yellow star thistle (*Centaurea solstitialis*) may increase thiamin need (Roberts et al., 1949; Lott, 1951; Martin, 1975). Caffeic acid (3,4-dihydroxycinnamic acid), a thiamin inhibitor isolated from bracken fern and a number of dicotyledonous seeds (Heimann et al., 1971), inhibits rat intestinal thiamin transport in vitro (Schaller and Holler, 1976). Bertone et al. (1984) gave 300 mg of caffeic acid orally or via a cecal fistula daily for 28 days to ponies fed a diet containing 0.9 mg of thiamin/kg of dry matter and reported significant increases in serum pyruvate and lactate concentrations compared to controls fed no caffeic acid. No estimates were made of proportional changes in quantitative thiamin requirement. Konishi and Ichijo (1984) reported muscular stiffness, decreased blood and urine thiamin concentrations, and increased blood pyruvate and lactate in two horses given orally a thermostable extract equivalent to 1.1 to 8.1 g of dried bracken fern/kg of body weight. Diniz et al. (1984) observed incoordination, staggering, and muscular tremors in twenty-seven mules, 2 months after introduction to bracken-infested pasture. Eight died and the rest recovered after removal from the pasture and injection with 100 mg of thiamin. At necropsy, generalized congestion, pulmonary edema, and serosal and mucosal hemorrhages were noted.

Amprolium (a coccidostat approved for use in poultry and cattle), which interferes with thiamin phosphorylation and membrane transport (Bauchop and King, 1968; Menon and Sognen, 1971), also increases dietary thiamin need (Cymbaluk et al., 1978), but the quantitative relationship between amprolium and thiamin has not been established. Signs of thiamin deficiency were induced in horses fed a diet containing 1.65 mg of thiamin/kg and dosed daily with 400 to 800 mg of amprolium/kg of body weight.

Although present data are limited, based on research with horses and other species the thiamin requirement appears to be no more than 3 mg/kg of dietary dry matter for maintenance, growth, and reproduction, unless high levels of antithiamin compounds are consumed. For performance horses, it may be prudent to ensure that their diets contain 5 mg of thiamin/kg of diet dry matter.

Signs of Deficiency or Excess   Experimentally produced thiamin deficiency caused anorexia, loss of weight, ataxia, bradycardia, missing heartbeats, muscular fasciculations, and periodic hypothermia of hooves, ears, and muzzle. Decreases in blood thiamin levels and erythrocyte transketolase activity and increases in blood pyruvate and lactate concentrations also occurred (Carroll et al., 1949; Cymbaluk et al., 1978; Bertone et al., 1984).

Oral toxicity is very unlikely for thiamin. The NRC (1987) notes that dietary intakes up to 1,000 times the requirement appear to be safe. Mackay (1961) administered thiamin to horses orally in dosages from 500 to 2,000 mg without observable effects. However, when thiamin was administered in doses of 1,000 to 2,000 mg by intramuscular or intravenous injection, a slowing of the pulse rate and a tranquilizing effect were reported. Irvine and Prentice (1962) criticized this work because of inadequate controls and reported a slowed heart rate also after saline injections. They attributed this effect to a transitory elevation of heart rate at the start of the experiment. Stewart (1972), likewise, was unable to demonstrate any significant effects on heart or respiratory rate from intravenous administration of 5 mg of thiamin hydrochloride/kg of body weight, either before or after exercise. On three occasions, horses injected with thiamin appeared to be less excitable while walking to the racecourse and did not exhibit the gross signs of tranquilization reported when they were injected with promazine. When thiamin is given parenterally in a single dose to dogs, mice, or rabbits, acutely toxic levels range from 50 to 400 mg/kg of body weight. The excess thiamin appears to block nerve transmission, producing curarelike signs including restlessness, epileptiform convulsions, cyanosis, and labored breathing. Death results from respiratory paralysis, usually accompanied by cardiac failure (Haley and Flesher, 1946).

RIBOFLAVIN

Riboflavin synthesis in the intestine of the adult horse or pony has been demonstrated by Jones et al. (1946), Carroll et al. (1949), and Linerode (1966). When Carroll et al. (1949) fed a diet containing 0.4 mg of riboflavin/kg of dry matter, riboflavin concentrations (mg/kg of ingesta dry matter) in the various intestinal sections were as follows: duodenum, 3.8; ileum, 1.1; cecum, 7.0; anterior large colon, 9.2; and anterior small colon, 12.2. Signs of riboflavin deficiency have not been described in horses or ponies fed diets providing 0.4 (Carroll et al., 1949), 2.2 (Jones et al., 1946), or 6.6 (Pearson and Schmidt, 1948) mg of riboflavin/kg of air-dry feed, although Pearson et al. (1944b) concluded that 2.2 mg of riboflavin/kg of air-dry feed is the approximate requirement for maintenance. The dietary riboflavin requirement is probably no more than 2 mg/kg of diet dry matter.

Signs of Deficiency or Excess Although riboflavin deficiency has not been described in horses, signs in other species include rough hair coat; atrophy of the epidermis, hair follicles, and sebaceous glands; dermatitis; vascularization of the cornea; catarrhal conjuctivitis; photophobia; and excess lacrimation. Some years ago, it was suggested that periodic ophthalmia (recurrent uveitis or moon blindness) is a consequence of riboflavin deficiency (Jones, 1942; Jones et al., 1945). However, the linkage between the two is not substantial, and invasions of the cornea by leptospira (Roberts, 1958) or microfilaria *(Onchocerca cervicalis)* (Cello, 1962) have been implicated in the production of periodic ophthalmia.

Little evidence exists of oral toxicity of riboflavin in any species. Schumacher et al. (1965) reported a reduction in pups born to rats supplemented with 104 mg of riboflavin/kg of diet. Estimates of the rat LD_{50} for intraperitoneal, subcutaneous, and oral administration are 0.56, 5, and more than 10 g of riboflavin/kg of body weight, respectively.

NIACIN

Niacin is a generic term for two compounds that have equal vitamin activity, nicotinic acid and nicotinamide. Pearson and Luecke (1944) observed that horses consuming 100 μg of nicotinic acid/kg of body weight excreted more nicotinic acid than had been ingested. Urinary excretion of this vitamin was influenced only slightly by dietary nicotinic acid intake. Niacin can probably be synthesized from tryptophan in the horse's tissues (Schweigert et al., 1947). Microbial synthesis in the horse intestine can be inferred from the observations of Carroll et al. (1949), who fed a diet containing 3 mg of nicotinic acid/kg of dry matter and found the following nicotinic acid concentrations (mg/kg of dry matter) in ingesta: duodenum, 55; ileum, 58; cecum, 121; anterior large colon, 96; and anterior small colon, 119. Linerode (1966) also concluded that appreciable microbial niacin synthesis occurs in the cecum and colon of the adult pony. No dietary niacin requirement has been established.

Signs of Deficiency or Excess Niacin deficiency has not been described in the horse. In other species, inflammation of lingual and buccal surfaces is common, followed by necrosis, ulceration, and foul breath odor. Dermatitis and hemorrhagic enteritis may also be seen.

Effects of niacin excess have not been described in the horse. However, high oral intakes of nicotinic acid have produced vasodilation, itching, sensations of heat, nausea, vomiting, headaches, and occasional skin lesions in humans (Robie, 1967; Hawkins, 1968). In addition, Winter and Boyer (1973) reported a hepatotoxicity from high intake of nicotinamide. Research with laboratory animals suggests that daily oral intake greater than 350 mg of nicotinic acid equivalents/kg of body weight can be toxic (NRC, 1987). Nicotinic acid may be tolerated somewhat better than nicotinamide. Limits for parenteral administration could be lower than those for oral intake.

PANTOTHENIC ACID

Pantothenic acid appears to be synthesized in the intestinal tract of adult horses (Carroll et al., 1949) and ponies (Linerode, 1966). Carroll et al. (1949) fed a diet containing 0.8 mg of panthothenic acid/kg of dry matter and found the following pantothenic acid concentrations (mg/kg of dry matter) in ingesta: duodenum, 11.7; ileum, 9.2; cecum, 39.2; anterior large colon, 34.4; and anterior small colon, 20.5. No signs of pantothenic acid deficiency were observed on this diet or on a second diet containing less than 0.2 mg of pantothenic acid/kg of dry matter. Likewise, no signs of deficiency were observed in growing ponies by Pearson and Schmidt (1948) who fed a diet containing 3.2 mg of pantothenic acid/kg of air-dry feed. No dietary pantothenic acid requirement has been established.

Signs of Deficiency or Excess Pantothenic acid deficiency has not been described in the horse. In other species, dermatitis, achromotrichia, enteritis, and neuritis have been noted. Degenerative changes in the peripheral motor nerves impair adductor muscle function in particular.

Ingestion of elevated dietary levels of pantothenic acid has produced no reported adverse reactions in any species (NRC, 1987). Feeding approximately 20 g of calcium pantothenate/kg of diet for 190 days to rats did not affect growth or gross pathology (Unna and Greslin, 1941). However, the acute LD_{50} for calcium pantothenate administered parenterally is about 1 g/kg of body weight.

VITAMIN B$_6$

Vitamin B$_6$ is a generic term for three metabolically interconvertible compounds: pyridoxine, pyridoxal, and pyridoxamine, which have equal vitamin activity on a molar basis. Pyridoxine appears to be synthesized by microorganisms in the intestinal tract of horses (Carroll et al., 1949). A diet containing less than 0.2 mg of pyridoxine/kg of dry matter resulted in the following pyridoxine concentrations (mg/kg of dry ingesta): duodenum, 2.2; ileum, less than 1.0; cecum, 2.4; anterior large colon, 6.1; and anterior small colon, 6.2. No dietary vitamin B₆ requirement has been established.

Signs of Deficiency or Excess   Vitamin B₆ deficiency has not been described in the horse. Signs in other species include scaly dermatitis; alopecia; microcytic, hypochromic anemia; impaired immune function; and neurological abnormalities resembling epilepsy.

Information is inadequate to estimate maximum tolerable dietary levels for species other than the rat and the dog, although oral doses of 2 to 6 g/day over 3 to 40 months have produced sensory nervous system dysfunction and disablement in adult humans (Schaumberg et al., 1983). Toxic signs include incoordination, ataxia, and convulsions, as well as bilateral loss of myelin and axons in the dorsal funiculi and loss of myelin in individual fibers of dorsal nerve roots. Krinke et al. (1980) concluded that the toxicity produces a peripheral, sensory neuropathy involving degeneration of the dorsal root ganglia, gasserian ganglia, and sensory nerve fibers. When administered for more than 60 days, 250 to 500 mg of pyridoxine/kg of diet can be tolerated by the rat; 500 to 1,000 mg/kg of diet can be tolerated for less than ·0 days.

BIOTIN

Biotin appears to be synthesized by microorganisms in the intestinal tract of the adult horse. Carroll et al. (1949) fed a diet containing less than 0.01 mg of biotin/kg of dry matter and found the following biotin concentrations in ingesta (mg/kg of dry matter): duodenum, less than 0.1; ileum, 0.1; cecum, 0.2; anterior large colon, 3.8; and anterior small colon, 2.3. No controlled studies have been published establishing a dietary biotin requirement above that supplied by intestinal synthesis. Clinical reports claim an improvement in hoof quality from feeding 10 to 30 mg of biotin/day for at least 6 to 9 months to adult horses (Comben et al., 1984). Varying degrees of improvement in hardness, integrity, and conformation of the hoof horn were reported. Kempson (1987) suggested that structural defects in the stratum externum of the hoof horn could be remedied by biotin alone, but poor attachment of the horn squames required biotin plus supplemental calcium and protein. Although the availability of biotin in barley, sorghum grain, and wheat appears to be lower than that in corn for poultry and swine (Anderson et al., 1978), biotin

availability in these grains has not been established for the horse.

Signs of Deficiency or Excess   No unequivocal evidence of biotin deficiency in the horse has been published. Signs in other species include inflammation and cracks on the plantar surface of the feet (Cunha et al., 1946, 1948).

Effects of excess biotin have not been described in the horse. Fetal resorption has been reported in rats injected subcutaneously with 50 to 100 mg of biotin/kg of body weight. Poultry and swine can tolerate at least 4 to 10 times their dietary requirement and probably much more (NRC, 1987).

FOLACIN

Folacin is the generic term for folic acid (pteroylmonoglutamic acid) and related compounds with similar biological activity. Folic acid appears to be synthesized by microorganisms in the intestine of the adult horse. Carroll et al. (1949) fed a diet containing less than 0.1 mg of folic acid/kg of dry matter and found the following folic acid concentrations in ingesta (mg/kg of dry matter): duodenum, 0.9; ileum, 0.5; cecum, 3.0; anterior large colon, 4.7; and anterior small colon, 2.7. Seckington et al. (1967) found lower serum folate levels in stabled than in grass-fed horses, but no estimates of folacin intake were provided. One stabled gelding in poor condition that performed poorly had a hemoglobin concentration of 11.5 g/dl and a serum folate value of 5 ng/ml. Daily oral administration of 20 mg of folic acid for 23 days was accompanied by an increase in concentrations of hemoglobin and serum folate to 14.6 g/dl and 12 ng/ml, respectively. Condition and performance also improved. Bone marrow biopsies revealed no evidence of megaloblastic anemia before treatment, although supplementary folic acid reportedly changed the appearance of the nucleus of red cell precursors. Allen (1978) used serum folate concentrations to assess the folate status of Thoroughbreds training to race, in-foal Thoroughbred mares, and crossbred ponies kept permanently on grass. The Thoroughbreds in training, although exhibiting no hematological or clinical abnormalities, had lower serum folate levels (mean 3.3, range 1.5 to 6.1 ng/ml) than in-foal mares (mean 10.6, range 6.4 to 15.8 ng/ml) or crossbred ponies on grass (mean 10.9, range 7.4 to 16.6 ng/ml). Intensive exercise over a 6-month period significantly decreased serum folate levels. Roberts (1983) also examined serum and erythrocyte folate levels in Australian horses and found higher concentrations in those on pasture than in permanently stabled horses. Because intramuscular injections of folic acid did not increase serum or erythrocyte values above normal beyond 24 hours postinjection, Roberts (1983) concluded that if folacin supplements were to be given, daily oral administration would be the method of choice. In contrast, Allen (1984) found that orally administered folic acid was absorbed poorly in the horse. This worker noted that natural folacin compounds are polyglutamates rather than monoglutamates, which may be absorbed by a different mechanism. Total folacin concentrations found in feedstuffs (μg/kg of wet weight) were as follows: oats, 210; hay, 670; wheat bran, 650; and fresh grass, 1,630. Thoroughbred racehorses were found to consume 6.3 mg of folacin/day, whereas grass-fed horses consumed 53.8 mg/day.

Signs of Deficiency or Excess Folacin deficiency has not been described in the horse. In some other species, a megaloblastic anemia with macrocytosis is seen.

No adverse responses to ingestion of excess folic acid have been reported in any species (NRC, 1987). However, single parenteral doses, about 1,000 times greater than the dietary requirement, have been reported to induce epileptic convulsions and renal hypertrophy in the rat.

VITAMIN B_{12}

Vitamin B_{12} is a generic term for corrinoids that exhibit qualitatively the biological activity of cyanocobalamin. High concentrations of vitamin B_{12} were found by Alexander and Davies (1969) in the contents of the large intestine of the horse. Davies (1971) subsequently demonstrated increased vitamin B_{12} concentration in gut contents from stomach to rectum. Values (ng/ml) were as follows: stomach, 4; small intestine, 5; cecum, 30; ventral colon, 84; dorsal colon, 197; and small colon, 301. These increasing concentrations are presumably the consequence of synthesis by intestinal anaerobes. Of 97 gut microbe isolates, 47 percent were found to produce vitamin B_{12} in vitro.

Stillions et al. (1971b) fed adult horses a semipurified diet containing about 1 μg of vitamin B_{12} and about 5 mg of cobalt/kg of air-dry feed. Although serum vitamin B_{12} concentration and daily urinary vitamin B_{12} excretion were lower than with a diet containing 90 μg of vitamin B_{12}/kg, daily vitamin B_{12} excretion was about 500 μg or five times greater than intake on the low- or high-vitamin B_{12} diet, respectively. No evidence of vitamin B_{12} deficiency was seen, and hemoglobin and hematocrit values were normal over an experimental period of 11 months. The same result is likely even with lower cobalt intake, because dietary cobalt levels greatly exceeded the theoretical need for vitamin B_{12} synthesis. In addi-

tion, horses have remained in good health while grazing pastures so low in cobalt that ruminants confined to them have died (Filmer, 1933). Evidence of absorption of vitamin B_{12} from the cecum and colon has been reported by Stillions et al. (1971b) and Salminen (1975). Caple et al. (1982) examined several hundred horses and reported plasma vitamin B_{12} concentrations of 1.8 to 7.3 μg/liter. Intramuscular injections of vitamin B_{12} were cleared rapidly from the plasma, and large amounts were excreted in the feces via the bile when vitamin B_{12} was administered intravenously in foals. Colostrum contributed significantly to the vitamin B_{12} status of the foal during the first 24 h after birth. Much of the vitamin was stored in the liver. Roberts (1983), in a survey of 88 horses in various states of physiology and training, found no evidence of vitamin B_{12} deficiency based on serum vitamin B_{12} concentrations or cellular hematology. No evidence of a dietary vitamin B_{12} requirement above that supplied by intestinal synthesis has been reported.

Signs of Deficiency or Excess Vitamin B_{12} deficiency has not been described in the horse. In some other species, normocytic, normochromic anemia and neurological abnormalities have been reported.

No unequivocal evidence of vitamin B_{12} toxicity in any species has been published (NRC, 1987). Winter and Mushett (1951) administered 1.6 g of vitamin B_{12}/kg of body weight intraperitoneally or intravenously (an amount several hundred times the dietary requirement) to mice and reported no adverse effects on growth or survival.

ASCORBIC ACID

Despite early suggestions that supplementary ascorbic acid (vitamin C) improved the sperm quality of stallions and the breeding performance of mares (Davis and Cole, 1943), other workers have been unable to repeat those results. Pearson et al. (1943) and Stillions et al. (1971a) suggested that tissues of the horse can synthesize ascorbic acid, and except for very large oral doses of the vitamin, dietary supplies have little relation to plasma and urinary ascorbic acid concentrations. Loscher et al. (1984) injected 5 or 10 g of ascorbic acid intravenously in horses and found that the distribution volume was approximately equivalent to the total body water. The terminal half-life of serum ascorbic acid was 5 to 17 h. Distribution and elimination correlated positively with dosage. After subcutaneous and intramuscular injection, the average bioavailability was 82 and 61 percent, respectively. Both routes of administration resulted in marked local irritation. Administration of 10 or 20 g of ascorbic acid in drinking water generally did not increase serum ascorbic acid concentrations above preadministration levels. These observations support the findings of Errington et al. (1954) and Jaeschke and Keller (1982) who also reported that ascorbic acid absorption from oral doses given to horses is very low. However, Snow and Frigg (1987a) reported that 20 g of oral ascorbic acid/day resulted in increased plasma ascorbic acid concentrations in horses in training, and Snow and Frigg (1987b) suggested that ascorbyl palmitate administered orally produced higher plasma ascorbic acid concentrations than ascorbic acid.

Signs of Excess Although ascorbic acid toxicity has not been reported in the horse, a number of toxicity symptoms and signs have been attributed to large intakes of ascorbic acid by humans and laboratory animals (NRC, 1987). These include oxaluria, uricosuria, hypoglycemia, excessive absorption of iron, diarrhea, allergic responses, destruction of vitamin B_{12}, and interference with hepatic mixed-function oxidase systems. However, some of these abnormalities were incidental and were noted in uncontrolled studies or are controversial. Mink seem to be quite sensitive to excess ascorbic acid, and 100 to 200 mg/kg of body weight/day produced a pronounced anemia in pregnant females and reduced the number and size of kits (Ender and Helgebostad, 1972). A dietary ascorbic acid concentration of 1 g/kg appears to pose no hazard to chickens, pigs, dogs, cats, and probably, horses.

CHOLINE

No studies of the choline requirement of the horse have been reported. Although not a vitamin in the strictest sense, choline can serve as a source of labile methyl groups for other species such as the rat, chicken, pig, dog, and cat. However, if methionine is available in excess of its specific dietary requirement, this amino acid can totally replace choline on an isomethyl basis.

Reports of the effects of excess choline intake on other species indicate that chickens and dogs should not be fed more than 2,000 mg of choline chloride/kg of dry diet (NRC, 1987). Rats, mice, and pigs appear to be unaffected by somewhat higher levels. The sensitivity of the horse to excess choline intake has not been established.

WATER

An adequate supply of clean water is essential for horses. The water content of the body is relatively constant (68 to 72 percent of the total weight on a fat-free basis) and cannot change appreciably without severe consequences for the horse. The minimal water requirement for any horse is the sum of the water lost from the body plus a component for growth in young animals (Robinson and McCance, 1952; Mitchell, 1962). Water is lost from the body in excretion of urine, feces, and sweat; in evaporation from the lungs and skin (insensible water loss); and through productive secretions such as milk. Lactation may increase the water need by 50 to 70 percent above that required for maintenance. In cases of foal diarrhea, fecal water losses can be substantial and may represent a serious health hazard for the foal. Also, as horses are produced and maintained for their athletic ability, the influence of exercise on water excretion is especially important. Water deprivation can lead to digestive disturbances such as colic (Argenzio et al., 1974).

One of the most important factors influencing water consumption is the dry matter intake. In an extensive review of the water requirements of animals, Leitch and Thomson (1944) concluded that horses need 2 to 3 liters of water/kg of dry matter intake. Fonnesbeck (1968) observed ad libitum water intake during metabolism trials with horses. Regression equations computed from the diet means showed that water intake could be estimated with a high degree of confidence from dry matter intake for horses receiving a maintenance diet. The trials demonstrated that all-hay diets resulted in a water-to-feed ratio of 3.6:1, whereas that of hay–grain diets was 2.9:1. An increase in environmental temperature may also increase the need for water very sharply. According to Lewis (1982), at $-18°C$, horses consume 2 l of water/kg of dry feed, whereas at 38C, they consume 8 liters/kg.

Work may increase the need for water as much as 20 to 300 percent (Carlson and Mansmann, 1974; Rose et al., 1977; Lucke and Hall, 1978; Snow et al., 1982). In the working horse, significant amounts of water are required for thermoregulation, and sweat losses from horses during extended periods of exercise can be substantial. According to Caljuk (1961), work can increase water needs to more than double, with evaporative losses increasing to almost one-third of intake. Morrison (1937) cited data obtained from a horse that gave off 9.4 kg of water vapor when trotting, an amount nearly twice that given off when walking and more than three times that when resting. Carlson (1982) suggested that change in body weight provides the most accurate means of estimating water losses. Losses of 5 to 10 percent of body weight have been found among horses during endurance rides (White et al., 1978; Snow et al., 1982).

The extent to which dehydration affects horse performance is not known, but water losses equivalent to 2 to 4 percent of body weight adversely affect performance in the human athlete (Saltin, 1964). Carlson (1982) noted that fluid and electrolyte disorders are major factors in equine exhaustion syndrome.

Dehydration and Electrolyte Balance

Dehydration through sweating results in the loss of both water and electrolytes. Snow et al. (1982) measured a mean total water loss of 39.8 liters of water, with a calculated loss of 6 liters from the plasma, from horses on an 80-km endurance ride. Sodium and chloride are the principal electrolytes present in equine sweat. A significant but lower amount of potassium is also present. Perspiration losses of these electrolytes may be very significant during extended work in hot, dry weather (Meyer et al., 1978; Carlson and Ocen, 1979; Frape et al., 1979; Rose et al., 1980; Snow et al., 1982). Concentrations (mmol/liter of sweat) range from 132 to 249 for sodium, 165 to 301 for chloride, and 32 to 78 for potassium. Electrolyte concentrations in sweat have been shown to be much higher than those in plasma (Rose et al., 1980; Snow et al., 1982). In heavy equine sweating, the loss of chloride can result in hypochloremia and metabolic alkalosis. The concentration of protein in equine sweat may be relatively high at the onset of sweating, but it declines to much lower levels during prolonged work. Total protein losses in sweat are relatively small (Kerr and Snow, 1983) and not considered a problem when horses are fed adequate amounts of normal energy sources (Hintz, 1983).

Replacement or preventive electrolyte therapy by oral or intravenous administration is of interest to horsemen. Intravenous administration of 1 to 3 liters of an electrolyte supplement is a regular part of prerace management among some trainers; however, no conclusive evidence exists that such treatment is beneficial (Carlson, 1982). An adequate water supply, a balanced diet, and free-choice trace mineralized salt should provide all necessary fluid and electrolytes in most racing situations. In some instances in a hot or humid environment, oral electrolyte supplementation may be beneficial for heavily sweating endurance horses (Carlson, 1982). Such supplementation should provide sodium and chlorine in roughly equal proportions, while the proportion of potassium should be one-third to one-half that of sodium. Calcium and magnesium may be added in small amounts. When such supplementation is deemed advisable, the amount given should not be excessive. The objective should be to provide partial replenishment of the deficits until such time as complete replacement by voluntary consumption occurs (Carlson, 1982).

Water Quality

The quality of the water supplied to horses is very important. In many cases the water available to horses can have adverse or toxic effects. A high salt content may result in water being undesirable, or in some cases unfit, for consumption. In the United States, this condition is more prevalent in the arid areas of the western part of the country. In 1950, officers of the Department of Agriculture and Governmental Chemical Laboratories in Australia have recommended about 6,500 mg/liter as the upper safe limit for salts in waters for horses (NRC, 1974). Pollution of the water supply can also result from stagnant or runoff water containing disease organisms or from industrial waste. Such industrial wastes may contain elements that are toxic to animals. The NRC (1974) has published a list of the recommended upper limits for some potentially toxic substances in the drinking water of livestock and poultry (see Table 2-3). Puls (1988) has also reviewed recommended maximum mineral levels.

TABLE 2-3 Safe Upper Level of Some Mineral Elements in Drinking Water for Horses

Mineral Element	Safe Upper Limit (mg/liter)
Arsenic	0.2
Cadmium	0.05
Chromium	1.0
Cobalt	1.0
Copper	0.5
Fluoride	2.0
Lead	0.1
Mercury	0.01
Nickel	1.0
Nitrate nitrogen	100.0
Nitrite nitrogen	10.0
Vanadium	0.1
Zinc	25.0

SOURCE: Adapted from NRC (1974).

Physical Characteristics and Suitability of Feeds

3

FORAGES

Forages are those feedstuffs that consist of the above-ground portions of herbaceous plants that are grown to be fed to animals. The more common forms of forage are pasture, soilage (green chop), hay, silage, and haylage. For feeding horses, pasture and hay are, by far, the most important forms of forage employed (Templeton, 1979). Although used to some extent to feed horses, silage and haylage may result in digestive upset because of the possible presence of molds and other toxic substances.

Feeding quality of the different forages used by horses varies widely due to plant species differences, soil fertility, climatic effects (available moisture and prevailing temperatures), and perhaps most important of all, the stage of maturity at the time of harvest (Sotola, 1937; Darlington and Hershberger, 1968; Taylor and Templeton, 1976; Rohweder et al., 1978; Utley et al., 1978; Hintz, 1983). Forages fed to horses must be free of toxins and other substances detrimental to the horse (see pp. 35–36). Some of the commonly grown U.S. pastures and hays at times may contain such substances.

Pastures

Pasture grasses may be perennials or annuals, and the perennials can be classified, according to their growth characteristics, as cool-season or warm-season grasses. The more important cool-season grasses in horse production programs are Kentucky bluegrass, timothy, orchardgrass, ryegrass, smooth bromegrass, and tall fescue. The more commonly used warm-season species are Bermudagrass, Bahiagrass, pangola (digitgrass), and the bluestems (Templeton, 1979). The cool-season grasses generally produce herbage with higher nutritive value than the warm-season grasses, because the latter

ordinarily contain higher concentrations of cell walls and are generally lower in digestibility. Pasture species should be selected for their adaptability to the area, as well as their nutrient composition.

Summer- and winter-annual grasses also can be grown as pasture sources or alternatives, although Sudangrass or sorghum-Sudangrass hybrids, which are widely grown annual forage species, are not recommended for horse pastures (Lewis, 1982). Horses grazing these pastures, or eating them as haylage or silage, may develop an inflammation of the urinary tract known as cystitis, and nervous and reproductive disorders have also been reported (Prichard and Voss, 1967; Adams et al., 1969; Van Kampen, 1970). Other annuals that are used in some areas include pearl millet and the small grains such as wheat, oats, and rye. It should be noted, however, that rye forage used for feed or bedding is safer for horses if grazed or cut prior to reaching the flowering or seed-forming stage, because of possible infection with ergot sclerotia (caused by the fungus *Claviceps purpurea*), which can have serious consequences for horses (Cheeke and Shull, 1985).

The legumes, of which alfalfa and the clovers are most important, are higher in protein, vitamins, and some minerals than the grasses. Pasture mixtures of grasses and legumes are excellent and offer a number of advantages over grasses alone, including superior nutrient value and longer growing seasons. Because of its growth characteristics, white clover is the most widely used legume in horse pasture mixtures (Smith et al., 1986).

The lush spring growth of both grasses and legumes is high in protein, vitamins, and minerals on a dry matter basis (NRC, 1982). However, the high moisture content in the forages may prevent grazing horses from consuming enough to provide the energy to meet their nutritive needs, and supplemental feeding may be required.

32

Likewise, as pasture plants mature, the percentage of water, protein, and most minerals decreases, while that of fiber increases; therefore, some classes of horses may require supplemental feed.

Hay and Other Preserved Forages

The factors that affect the nutritive value of pasture crops have similar effects on hay and other preserved forages. Differences in methods of preservation and processing also influence the nutrient content of preserved forages. Good-quality grass hays are palatable to horses and may be fed as the only roughage. High-quality legume hays, such as alfalfa, provide higher levels of readily utilizable soluble carbohydrate, protein, calcium, and carotene than do most grass hays (Fonnesbeck and Symons, 1967; Fonnesbeck, 1969). Legume hays may also be fed as the only roughage. Mixtures of grass and legumes make excellent roughages for horses. Weathered, stemmy, or nutritionally deficient hays can be fed if adequate supplementation is provided. Moldy or dusty hay should not be used (Lewis, 1982; Hintz, 1983; Smith et al., 1986).

Grass hays most commonly fed to horses include timothy, orchardgrass, smooth bromegrass, coastal Bermudagrass, oat hay, prairie hay, tall fescue, and Sudangrass. These grass hays can provide excellent forage for horses if harvested before becoming overly mature, cured without weather damage, and stored properly. A recent report by J. C. Reagor (Texas Veterinary Medicine Diagnostic Laboratory, unpublished data, 1987) indicates that Sudangrass hay may cause the same types of problems described for horses consuming Sudangrass pasture. Good-quality grass hay should be leafy, soft, and pliable to the touch, have no or comparatively few seed heads, and should be free of mold, dust, and weeds. With oat hay, if the heads are lost the quality decreases. Riveland et al. (1977) have suggested that oat hay may contain toxic amount of nitrates and should be tested before feeding. If harvested for forage, oats should be cut when the grains are in the late-milk to early-dough stage (Sotola, 1937; Riveland et al., 1977; Ensminger and Olentine, 1978).

Legume hays most commonly fed to horses include alfalfa, red clover, birdsfoot trefoil, and to a lesser extent, lespedeza. Of these, alfalfa is generally highest in quality, being less stemmy and somewhat higher in protein and calcium content. Red clover, if properly harvested and stored, can have a quality similar to that of alfalfa, but it is often stemmy and lacking in bright, green color. Birdsfoot trefoil is usually less stemmy than red clover and may be similar in composition to that of alfalfa, but it is less widely used for horses (Hintz, 1983). In recent years some varieties of perennial peanuts (*Arachis* sp.) have been developed, which are promising forages for hay production on well-drained soils in Florida and other areas with similar climatic conditions (Prine, 1972; Woodroof, 1983).

The physical form of hay is of interest in feeding horses. Hintz (1983) reported that chopping hay did not influence digestibility or increase intake but, when fed mixed with grain, resulted in a slower rate of intake of the grain. Pelleted and cubed forages can also be fed successfully to horses (Haenlein et al., 1966; Hintz and Loy, 1966; Schurg et al.,1977; Lewis, 1982; Jackson et al., 1985; Cymbaluk and Christensen, 1986). In addition, both bred mares and mature geldings at light work can be maintained on a diet of alfalfa-whole corn plant (1:1) pellets plus alfalfa hay cubes, although considerable appetite depravity (wood chewing) is observed (J. P. Baker, University of Kentucky, unpublished data, 1978).

Most forages can be stored successfully as hay if allowed to dry to 20 percent or lower in moisture content. At higher moisture levels, the forage is subject to molds, heating, and spoilage. A serious outbreak of botulism, which resulted in the death of eight horses from eating improperly cured chopped oat hay, has been reported (Kelly et al., 1984).

If a forage is dried excessively, much of the leafy material is lost and the hay quality is seriously reduced. When a forage is harvested for hay at higher moisture levels, preservatives can be applied to prevent molding and spoilage. One preservative product consisting of propionic acid (80 percent) and acetic acid (20 percent) has been shown to successfully preserve alfalfa hay baled at moisture levels up to 35 percent, and this hay has been fed to horses with excellent results (Hintz et al., 1983; Lawrence et al., 1987; Battle et al., 1988).

With forages harvested at higher moisture levels and stored as haylage or silage, the risk of substances toxic to horses is much greater (Smith and Murray, 1984). A serious outbreak of botulism from horses eating big bale silage was reported by Ricketts et al. (1984). Essential to the successful preservation of such high-moisture forage material is the exclusion of air to produce anaerobic conditions and maintain a pH low enough to prevent mold development. The optimum pH ranges from 3.5 to 4.5. A patented procedure for vacuum packing high-moisture forage ("HorseHage") was developed in England and recently introduced to the United States; it is now being marketed to the horse industry.

Other Roughages

CROP RESIDUES

Straws from the cereal grains are much higher in fiber and lower in digestible energy, protein, calcium, and phosphorus than either grass or legume hays (Sotola, 1937; NRC, 1982), but they can be used in horse diets to provide fiber if accompanied by adequate supplemental feeds. Some types of straw processing have been shown to be beneficial. A number of studies have shown that straw treated with ammonia, sodium hydroxide, or acid followed by yeast inoculation is more digestible than untreated straw (Schurg, 1981; Hintz, 1983).

Ground corncobs, although very low in protein, essential minerals, and vitamins, provide approximately 50 to 70 percent as much DE as the average grass hays (Table 6–1A), and can add important fiber or bulk to horse diets.

BY-PRODUCT ROUGHAGES

When adequate quantities of high-quality forages are not available, by-product roughages may have to be used in horse diets. Cottonseed hulls are a by-product roughage used widely in cattle feeds for many years but used to only a limited extent in horse feeds. Cottonseed hulls are very low in digestible protein, calcium, phosphorus, and vitamins. Cottonseed hulls contain approximately 50 to 70 percent as much DE as average-quality grass hay (Table 6–1A). Despite their low nutritive value, cottonseed hulls can be used to a limited extent to provide fiber or bulk.

Peanut hulls, another by-product roughage, have rarely been fed to horses but might be used to provide additional fiber or bulk in horse diets, although peanuts are susceptible to invasion by aflatoxins (Cheeke and Shull, 1985). Peanut hulls are somewhat higher in crude protein than cottonseed hulls, but they contain only about two-thirds as much DE (Table 6–1A).

Rice mill feed, a by-product of rice milling, consists of parts of the rice grain and hull, but it must not contain more than 32 percent crude fiber. Where readily available, it is used in some horse feeds.

Another type of roughage that may find increasing use in horse feed is paper. With the vast amounts of paper products being used and then discarded, a good supply seems assured. In periods of extreme feed shortage in Europe during World War II, horses were fed paper as a source of roughage, and more recently, in recent research, they were fed pelleted diets containing 25 percent computer paper or ground corrugated paper boxes (Hintz, 1983). The paper products did not contain significant amounts of protein, essential minerals, or vitamins, but did provide significant amounts of digestible cellulose and the DE content was similar to that of alfalfa meal. Because many different inks are used in printing, consideration should be given to their potential toxicity if paper products are to be fed.

ENERGY FEEDS

Grains

Cereal grains are fed primarily as sources of energy, and all of the common grains may be used for horses. Differences in nutritive value and physical characteristics of grains should be considered when formulating diets. Oats, the traditional grain for horses, are higher in fiber, are lower in DE, and weigh less per unit volume than other grains. Oats can be fed whole, although rolling or crimping may improve their utilization by approximately 5 percent (Morrison et al., 1919). For young foals and for old horses with poor teeth, rolling or crimping of oats is advisable.

Corn is an excellent source of energy for horses. It is lower in protein and higher in DE than oats. It is almost invariably a more economical source of energy than oats. When replacing oats with corn in a diet, one should remember that a given volume of corn provides approximately twice as much DE as the same volume of oats.

Barley may be fed to horses as the only grain. Barley is higher in DE than oats, but lower than corn. It should be steam rolled or crimped. The outer hull adheres tightly to the kernel; thus, processing increases digestibility (Ott, 1972).

Sorghum grains and wheat may be fed to horses, but these small, dense grains should be rolled, cracked, or steam flaked to facilitate digestion (Klendshoj et al., 1980; Lewis, 1982). Wheat, if ground, tends to be gummy when eaten and, if used, should be steam rolled, mixed with some bulky feed, and included at not more than 20 percent of the concentrate mix (Cunha, 1980).

Rye can be fed to horses if it is not contaminated with ergot, but it is not as palatable as other cereals and should be limited to no more than 10 to 20 percent of the diet. If rye is fed to horses, it should be processed to improve digestibility.

By-Product Energy Feeds

There are a number of by-products of the food and feed industries that can supply significant amounts of the DE needed by horses. Such by-product feeds include beet pulp, citrus pulp, hominy feed, feeding oatmeal, oat mill by-products, rice polishings, soybean hulls, wheat bran, wheat middlings, wheat mill run, and dried whey (Hintz, 1983). The DE content of these feeds compares favorably to that of the feed grains, whereas their protein and mineral contents are usually somewhat higher than those of the feed grains (Table 6–1A).

The various fats and oils produced from rendering and extraction processes can be added in limited amounts to horse diets. They will reduce the dustiness and add significantly to the DE content of the feed (Bowman et al., 1977; Hintz et al., 1978; White et al., 1978; Kane et al., 1979; Hambleton et al., 1980; Rich et al., 1981; Duren et al., 1986). Molasses can also supply DE, reduce dustiness, prevent separation of fine materials in the feed, and improve palatability. By-products of the brewing and distilling industries, dried brewers grains, and dried distillers grains not only provide DE, minerals, and vitamins, but also add significant amounts of protein to the diet (Leonard et al., 1975; Cunha, 1980; Hintz, 1983; Webb et al, 1985; Frape, 1986). Dehydrated brewers yeast is added often to mixed feeds because it is a rich source of B vitamins (Webb et al., 1985).

PROTEIN SUPPLEMENTS

Plant Protein Supplements

Soybean meal, cottonseed meal, and linseed meal (flaxseed meal) are the most important sources of supplemental plant protein. Other plant protein sources that can be used include peanut meal, safflower meal, canola meal, sunflower meal, corn gluten meal, and sesame meal (Cunha, 1980; Lewis, 1982; Hintz, 1983; Frape, 1986).

Animal Protein Supplements

Casein, dried skim milk, fish meal, and meat meal are protein feeds of animal origin that may be used to feed horses (Cunha, 1980; Lewis, 1982; Hintz, 1983; Frape, 1986). Casein, fish meal, and meat meal are not widely used in horse feeds in the United States, due partly to their cost and availability. Meat meal is often unpalatable to horses. Many animal protein products contain considerable amounts of fat, which may result in rancidity (Lewis, 1982), and some fish meals may be contaminated with *Salmonella* or other enteric organisms that cause diarrhea (Frape, 1986). Dried skim milk is used in many foal feeds.

Fish meal and the milk protein sources have a protein quality superior to that of other protein sources because of their comparatively higher lysine concentration (Table 6–1A).

FEED PROCESSING

Hay is usually fed in loose form but may be processed into pellets, cubes, or wafers (Lewis, 1982; Hintz, 1983;

Cymbaluk and Christensen, 1986; Frape, 1986). Cereal grains can be fed whole or processed by cracking, rolling or crimping, steam flaking, micronizing, or extruding (Cunha, 1980; Lewis, 1982). Processing increases the digestibility of oats and barley by 2 to 5 percent; corn, 7 to 9 percent; and milo, wheat, and rye, approximately 15 percent (Klendshoj et al., 1980; Lewis, 1982). Fine grinding should be avoided unless the feed is subsequently pelleted or extruded. Pelleted feeds have several advantages such as reduced waste, dust, and storage space required (Hintz, 1983), but there are also disadvantages. Pelleted concentrate feeds, especially very small pellets, may be eaten quite rapidly and, because the feeds they contain have been finely ground before being pelleted, may be more prone to cause colic due to overconsumption and rapid fermentation in the horse's digestive tract. Increasing the fiber content of the pelleted concentrates may decrease the rate of feed intake and help minimize these problems. It is especially important to maintain a readily available supply of roughage for horses being given pelleted feeds (Haenlein et al., 1966; Willard et al., 1977). Integrity of the pellets and cubes is also important. If they are soft and crumbly, fine material will break off and be lost, or digestive problems may result (Lewis, 1982).

SOME FEED PROBLEMS

Examples of problems that can result from toxins and other substances sometimes found in certain horse feeds are described below. Additional information is available in *Interactions of Mycotoxins in Animal Production* (NRC, 1979).

Equine leucoencephalomalacia, a condition commonly known as moldy corn poisoning or cornstalk disease, results in sporadic loss of horses that have eaten moldy corn or corn forage. The causative organism is reported to be associated with a fungus *Fusarium moniliforme* (Ley, 1985).

Ergot, a fungus (*Claviceps* spp.) can infect certain grains and forage species and result in serious health problems in horses (Cheeke and Shull, 1985). *C. purpurea* may infect several species of ryegrass (*Lolium* spp.), canary grass (*Phalaris canariensis*), and a number of other native grasses found in the western United States. Ergotism from this fungus resulted in paralysis and death of horses in 6 to 12 h (Wilcox, 1899). In two cases, horses that had eaten large quantities of wild ryegrass infected with *C. purpurea* developed gangrene of the legs. *Claviceps paspali* may infect Dallis grass (*Paspalum dilatatum*) and Argentine Bahiagrass (*Paspalum notatum).* Simms (1951) reported that hay made from infected Dallis grass produced convulsions in horses.

Fescue toxicity, a condition caused by an endophyte fungus or a toxic substance within the tall fescue plant, may result in prolonged gestation periods; tough, thickened placentas; and agalactia in mares (Garrett et al., 1980; Harper and Henton, 1981; Barnett et al., 1984; Cheeke and Shull, 1985).

Oxalate poisoning results in nutritional secondary hyperparathyroidism ("bighead disease") and other skeletal problems in horses grazing certain tropical grasses, such as kikuyu *(Pennisetum clandestinum)* and setaria *(Setaria sphacelata)* (Jones et al., 1970; Cheeke and Shull, 1985). Ward et al. (1979) demonstrated that approximately 20 to 30 percent of the calcium in alfalfa is in the form of oxalate and not available to ruminants, and Cymbaluk et al. (1986) suggested that oxalate might bind 38 to 44 percent of the calcium in native Canadian grass hays *(Calamagrostis* and *Carex* spp.). However, Hintz et al. (1984) conducted an experiment with alfalfa samples taken from different areas in the United States and Canada and found no problems with mineral availability for ponies.

Excessive salivation ("slobbering") can result from a fungus *(Rhizoctonia leguminicola)* infection of red clover. It is most common during the warm, moist conditions that often prevail in the Southeast at the time of the second cutting (Aust et al., 1968; Sockett et al., 1982; Cheeke and Shull, 1985).

Alsike clover may cause toxic signs in horses, with photosensitization being the most common problem (Cheeke and Shull, 1985).

Hordenine, a substance present in certain feedstuffs such as sprouted barley and reed canarygrass, has a stimulating effect on the heart. Hordenine residue in the urine has resulted in the disqualification of some racehorses (Reilly, 1981; Cheeke and Shull, 1985).

Mycotoxins such as aflatoxins, toxic compounds that are produced by certain strains of *Aspergillus* spp., may contaminate corn, soybeans, cottonseed, or peanuts (Cheeke and Shull, 1985). Consumption of contaminated feeds can result in feed refusals and in serious liver and kidney damage.

Alfalfa hay, especially that grown in the central United States, can be seriously infested with blister beetles that have been killed and preserved in the hay. These beetles, which may be black, yellowish brown with dark stripes, or gray, contain cantharidin, an irritant that is severely toxic to horses that ingest it (Schoeb and Panciera, 1978; Beasley et al., 1983; Christensen, 1987). Alfalfa hay from that section of the country should be examined closely because consumption of a significant number of blister beetles can be fatal to horses.

4 General Considerations for Feeding Management

FEEDING MANAGEMENT

Selecting a Diet

The diet selected for different horses will depend on the nutrient requirements of particular classes of horses and the combinations of ingredients chosen to meet those requirements. Varying combinations of ingredients may be chosen depending on nutrient content, availability, price of ingredients, and preference of the horse owner. All diets for horses should contain adequate amounts of roughage. Thus, all diets must contain either pasture or some type of harvested roughage such as hay or other forms of roughage described in Chapter 3. Varying the proportions of roughage and concentrates can be used in the management of horses (1) to control energy intake, (2) to maintain normal digestive tract fill, (3) to minimize digestive dysfunction, and (4) to regulate consumption of feeds by horses that are fed in groups. Concentrate-to-roughage ratios, as suggested in Table 5-2, can be modified to suit individual situations. However, it is highly recommended that all horses either have access to a pasture or be fed sufficient long roughage in the diet to minimize the digestive dysfunction often attributed to feeding large amounts of concentrates. A good rule of thumb is that horses should be fed at least 1 percent of their body weight/day of good-quality roughage or be given access to a pasture for sufficient time to allow them to consume at least 1 percent of body weight of dry matter/day. Grazing horses and horses fed good-quality roughages ad libitum will voluntarily consume from 2 to 2.5 percent of their body weight as dry matter in a 24-h period. When energy requirements of the horse require that the diet contain more energy than can be supplied by roughages alone, the diet should contain at least 1 kg of roughage DM/100 kg of body weight and sufficient concentrates to meet the animal's energy requirements. Once the roughage has been selected, the concentrates must be balanced for all nutrients so that the final ration will meet the total daily nutrient requirements of the animal in an amount of feed the animal will consume. See Table 5-4 for expected intake of different types of horses.

Feeding Adult Horses

The energy requirements of mature horses at maintenance are low and can be met by feeding good-quality roughages. The only supplemental feed required will probably be salt and a balanced mineral supplement provided free-choice. Other approaches can be taken with the mature, idle horse, including using some concentrate feeds in a restricted-roughage feeding management program.

Feeding management for broodmares requires judicious attention to the maintenance of adequate body condition. Broodmares in late pregnancy should be fed good-quality roughages, or a combination of roughages and concentrates, in sufficient amounts to allow the mare to store body fat that can be utilized for energy needs during early lactation and rebreeding. Mares fed average- or poor-quality roughages during early lactation must be fed balanced concentrates in sufficient amounts to meet energy and other nutrient needs.

Because it is difficult for a horse to consume sufficient roughage to meet the energy needed for hard work, concentrates are normally required to meet working horses' energy needs. Thus, the concentrate-to-roughage ratio in diets for working horses should increase as the intensity or amount of work increases. However, even the working horse needs some long roughage in the diet to maintain normal intestinal health. Working horses are best fed individually, rather than in groups, to avoid feeding competition that could result in some horses being underfed and others being overfed.

Feeding Growing Horses

The nutrient needs of the young growing horse must be met to achieve optimum growth and development and structural soundness at maturity. The most critical concept in feeding the young growing horse is to feed a *balanced diet*. A balanced diet is defined as one in which all nutrients are supplied in adequate amounts, but just as important, they are supplied in correct amounts relative to each other. It has been suggested that dietary imbalances may be the causative factors in a variety of developmental orthopedic diseases in young horses. Therefore, diets for young horses must be balanced carefully and fed in the amounts necessary to meet all nutrient requirements proportionate to energy consumption. Tables 5-1 and 5-2 list two levels of nutrient intake for growing horses, one for moderate growth and one for more rapid growth.

Creep Feeding of Foals

In many situations, it is desirable to give suckling foals an appropriate supplemental feed prior to weaning. Mares that are poor milkers, or have impaired production due to disease or other factors, may not produce adequate nutrients for the foal during early lactation. Furthermore, nutrient secretion by the typical mare may not be adequate for the foal after it reaches 3 months of age. One of the most workable practices to ensure adequate nutrient intake for suckling foals is the use of a creep feeder. The creep feeder should be strategically located near places that mares frequent, such as watering or feeding areas, and should be designed to allow the foal access to the feed in a safe manner without injury. Supplemental feed for suckling foals should be provided at least once daily, more often if weather and other conditions indicate. Feed should be provided in liberal quantities so that all foals have free-choice access to feed any time they want to eat. Pelleted creep feeds may be preferable to textured feeds because they prevent sorting of ingredients. One of the most important advantages of creep feeding is to accustom foals to eating concentrates before they are weaned. Foals that have been provided with creep feed for a period of several weeks prior to weaning generally eat better after weaning, and may be less susceptible to the stresses of weaning than foals that have not been so fed. Providing supplemental feed in a creep feeder for foals is preferable to having the foal eat the dam's concentrate because the foal's nutrient requirements relative to its energy needs are higher than those of the mare.

FEED ADDITIVES AND OTHER COMPOUNDS

Although the practice of using additives such as antibiotics, growth stimulants, and various drugs is commonplace in other livestock feeds, very little research has been done on the value of these compounds in horse feeds. There have been field reports of short-term benefit from feeding antibiotics to horses to help prevent infectious disease under stressful situations, but their long-term effects are not known. Until sufficient information is available on the efficacy of these additives, the horse owner is advised to proceed with caution when considering the use of antibiotics or other drugs in the diets of horses.

Anabolic steroids are being used in several situations in the horse industry, but many of these uses are not warranted. Although anabolic steroids may enhance appetite and assist in rehabilitating horses in a debilitated state (O'Conner et al., 1973), no evidence exists that they are beneficial to healthy, well-fed horses (Burke et al., 1981). Furthermore, detrimental effects can arise from the use of anabolic steroids in young horses. Anabolic steroids cause testicular degeneration in young males (Squires et al., 1982) and reproductive dysfunction in young females (Squires et al., 1985). Therefore, the use of anabolic steroids in the diet or in the management of young, healthy, growing horses is not recommended.

Some growth-promoting agents used in other livestock feeds are toxic to the horse. For example, monensin is an ionophore that has beneficial effects for cattle and poultry. However, monensin can be lethal to horses if ingested in amounts that would be safe for cattle and poultry (Matsuaka, 1976). Therefore, any feed designed for use by other species should be evaluated carefully before being fed to horses.

5 Nutrient Requirement Tables

USING THE EQUATIONS AND TABLES

The equations provided in this chapter can be used to calculate the requirements for digestible energy, crude protein, lysine, calcium, phosphorus, magnesium, potassium, and vitamin A. Factors considered in the derivation of the equations are discussed in Chapter 2.

Table 5-1 presents the daily nutrient requirements calculated from the equations. These requirements are the total daily requirements that must be supplied in the sum of forage and concentrates that horses consume. In most cases the daily nutrient requirements are calculated from actual experimental data on horses, and some are based on the energy requirements of particular classes. For example, nutrient requirements for the working horses are given in a constant ratio of nutrient relative to DE consumption. In some cases, nutrient requirements may be liberal. However, to calculate values in Table 5-1, it was assumed that as the DE requirement increases, a comparable increase occurs in the requirements of other nutrients for the working horse.

The nutrient concentrations given in Tables 5-2 and 5-3 reflect the requirements in the total diet of horses, including roughages and concentrates. The nutrient concentrations in Table 5-2 are accurate only when the energy density specified for each classification of horse is used because the concentration allowances are calculated from the daily nutrient requirements and the expected feed intake (Table 5-4). In many cases the nutrient concentrations shown in Table 5-2 are rounded for ease of application.

EQUATIONS FOR CALCULATING DAILY REQUIREMENTS OF HORSES*

I. Estimation of digestible energy (DE) requirements (Mcal of DE/day)†

    A. Maintenance:

        1. 200–600 kg of BW $DE = 1.4 + 0.03\,BW$

        2. Greater than 600 kg of BW $DE = 1.82 + 0.0383\,BW - 0.000015\,BW^2$

    B. Stallions (breeding season): $DE = 1.25(\text{maintenance DE})$

    C. Pregnant mares:

        1. 9 months $DE = 1.11(\text{maintenance DE})$

        2. 10 months $DE = 1.13(\text{maintenance DE})$

        3. 11 months $DE = 1.20(\text{maintenance DE})$

    D. Lactating mares:

        1. Foaling to 3 months

            a. 200 kg of BW $DE = (\text{maintenance DE}) + (0.04\,BW \times 0.792)$

            b. 400–900 kg of BW $DE = (\text{maintenance DE}) + (0.03\,BW \times 0.792)$

---

*Factors considered in the derivation of the equations are discussed in the text. All values are on a dry matter basis.

†BW = body weight (kg).

    2. 3 months to weaning
       a. 200 kg of BW   DE = (maintenance DE) + (0.03 BW × 0.792)
       b. 400–900 kg of BW   DE = (maintenance DE) + (0.02 BW × 0.792)

E. Working horses:
    1. Light work   DE = 1.25(maintenance DE)
    2. Moderate work   DE = 1.50(maintenance DE)
    3. Intense work   DE = 2.00(maintenance DE)

F. Growing horses (4–24 months of age):
    1. Not in training   DE = (maintenance DE) + $(4.81 + 1.17X - 0.023X^2)$(ADG)
    2. In training   DE = 1.5(maintenance DE) + $(4.81 + 1.17X - 0.023X^2)$(ADG)
      (moderate work)

where X is the age (months) and ADG is the average daily gain (kg/day).

II. Estimation of crude protein (CP) requirements (g/day)*
  A. Maintenance: CP = (40)(Mcal of DE/day)
  B. Stallion: CP = (40)(Mcal of DE/day)
  C. Pregnant mares
    (9–11 months):   CP = (44)(Mcal of DE/day)
  D. Lactating mares:
    1. Foaling to 3 months

      a. 200 kg of BW

$$CP = \frac{(\text{maintenance DP}) + [(0.04\,BW \times 0.021 \times 1{,}000)/0.65]}{0.55\dagger}$$

      b. 400–900 kg of BW

$$CP = \frac{(\text{maintenance DP}) + [(0.03\,BW \times 0.021 \times 1{,}000)/0.65]}{0.55\dagger}$$

    2. 3 months to weaning

      a. 200 kg of BW

$$CP = \frac{(\text{maintenance DP}) + [(0.03\,BW \times 0.018 \times 1{,}000)/0.65]}{0.55\dagger}$$

      b. 400–900 kg of BW

$$CP = \frac{(\text{maintenance DP}) + [(0.02\,BW \times 0.018 \times 1{,}000)/0.65]}{0.55\dagger}$$

  E. Working horses: CP = (40)(Mcal of DE/day)
  F. Growing horses:
    1. Weanlings   CP = (50)(Mcal of DE/day)
    2. Yearlings and long yearlings   CP = (45)(Mcal of DE/day)
    3. 2 year olds   CP = (42.5)(Mcal of DE/day)

III. Estimation of lysine requirements (g/day)
  A. Mature horses: lysine = (0.035)(g of CP/day)
  B. Weanlings: lysine = (2.1)(Mcal of DE/day)
  C. Yearlings and long yearlings: lysine = (1.9)(Mcal of DE/day)
  D. Two-year olds: lysine = (1.7)(Mcal of DE/day)

IV. Estimation of calcium (Ca) requirements (g/day)*
  A. Maintenance: Ca = 0.04 BW
  B. Stallions: Ca = (1.22)(Mcal of DE/day)
  C. Pregnant mares
    (9–11 months):   Ca = (1.90)(Mcal of DE/day)
  D. Lactating mares:
    1. Foaling to 3 months
      a. 200 kg of BW   Ca = (maintenance Ca) + [(0.04 BW × 1.2)/0.5]
      b. 400–900 kg of BW Ca = (maintenance Ca) + [(0.03 BW × 1.2)/0.5]
    2. 3 months to weaning
      a. 200 kg of BW   Ca = (maintenance Ca) + [(0.03 BW × 0.8)/0.5]
      b. 400–900 kg of BW Ca = (maintenance Ca) + [(0.02 BW × 0.8)/0.5]
  E. Working horses: Ca = (1.22)(Mcal of DE/day)

*BW = body weight (kg).
†Crude protein digestibility.

F. Growing horses
  1. Not in training   Ca = 0.04 BW + 32 ADG

  2. In training
      (moderate work)

$$Ca = \frac{\text{Ca required for horses not in training}}{\text{DE required for horses not in training}} \times \text{DE required for horses in training}$$

V. Estimation of phosphorus (P) requirements (g/day)*
  A. Maintenance: P = 0.028 BW
  B. Stallions: P = (0.87)(Mcal of DE/day)
  C. Pregnant mares
      (9–11 months):   P = (1.44)(Mcal of DE/day)
  D. Lactating mares:
    1. Foaling to 3 months

      a. 200 kg of BW $P = \dfrac{0.010\,BW + (0.04\,BW \times 0.75)}{0.45}$

      b. 300–900 kg of BW $P = \dfrac{0.010\,BW + (0.03\,BW \times 0.75)}{0.45}$

    2. 3 months to weaning

      a. 200 kg of BW $P = \dfrac{0.010\,BW + (0.03\,BW \times 0.50)}{0.45}$

      b. 300–900 kg of BW $P = \dfrac{0.010\,BW + (0.02\,BW \times 0.50)}{0.45}$

  E. Working horses: P = 0.87 (Mcal of DE/day)
  F. Growing horses:
    1. Not in training   P = 0.022 BW + 17.8 ADG

    2. In training
        (moderate work)

$$P = \frac{\text{P required for horses not in training}}{\text{DE required for horses not in training}} \times \text{DE required for horses in training}$$

VI. Estimation of magnesium (Mg) requirements (g/day)*
  A. Maintenance: Mg = 0.015 BW
  B. Stallions: Mg = 0.46 (Mcal of DE/day)
  C. Pregnant mares
      (9–11 months):   Mg = 0.48 (Mcal of DE/day)
  D. Lactating mares:
    1. Foaling to 3 months
      a. 200 kg of BW   Mg = (maintenance Mg) + [(0.04 BW × 0.09)/0.4]
      b. 400–900 kg of BW   Mg = (maintenance Mg) + [(0.03 BW × 0.09)/0.4]
    2. 3 months to weaning
      a. 200 kg of BW   Mg = (maintenance Mg) + [(0.03 BW × 0.045)/0.4]
      b. 400–900 kg of BW   Mg = (maintenance Mg) + [(0.02 BW × 0.045)/0.4]
  E. Working horses: Mg = 0.46 (Mcal of DE/day)
  F. Growing horses:
    1. Not in training   Mg = 0.015 BW + 1.25 ADG

    2. In training
        (moderate work)

$$Mg = \frac{\text{Mg required for horses not in training}}{\text{DE required for horses not in training}} \times \text{DE required for horses in training}$$

---

*BW = body weight (kg).

VII. Estimation of potassium (K) requirements (g/day)*
    A. Maintenance: $K = 0.05\,BW$
    B. Stallions: $K = 1.52\,(\text{Mcal of DE/day})$
    C. Pregnant mares
       (9–11 months): $K = 1.60\,(\text{Mcal of DE/day})$
    D. Lactating mares:
       1. Foaling to 3 months
          a. 200 kg of BW $K = (\text{maintenance K}) + [(0.04\,BW \times 0.7)/0.5]$
          b. 400–900 kg of BW $K = (\text{maintenance K}) + [(0.03\,BW \times 0.7)/0.5]$
       2. 3 months to weaning
          a. 200 kg of BW $K = (\text{maintenance K}) + [(0.03\,BW \times 0.4)/0.5]$
          b. 400–900 kg of BW $K = (\text{maintenance K}) + [(0.02\,BW \times 0.4)/0.5]$
    E. Working horses: $K = 1.52\,(\text{Mcal/day})$
    F. Growing horses:
       1. Not in training $K = (0.05\,BW) + (3.0\,ADG)$

       2. In training
         (moderate work)

$$K = \frac{K \text{ required for horses not in training}}{DE \text{ required for horses not in training}} \times DE \text{ required for horses in training}$$

VIII. Estimation of vitamin A requirements (IU/day)*
    A. Maintenance: vitamin A $= 30\,BW$
    B. Pregnant and lactating mares: vitamin A $= 60\,BW$
    C. All others: vitamin A $= 45\,BW$

*BW = body weight (kg).

TABLE 5-1A   Daily Nutrient Requirements of Ponies (200-kg mature weight)

Animal	Weight (kg)	Daily Gain (kg)	DE (Mcal)	Crude Protein (g)	Lysine (g)	Calcium (g)	Phosphorus (g)	Magnesium (g)	Potassium (g)	Vitamin A (10^3 IU)
Mature horses										
Maintenance	200		7.4	296	10	8	6	3.0	10.0	6
Stallions	200		9.3	370	13	11	8	4.3	14.1	9
(breeding season)										
Pregnant mares										
9 months	200		8.2	361	13	16	12	3.9	13.1	12
10 months			8.4	368	13	16	12	4.0	13.4	12
11 months			8.9	391	14	17	13	4.3	14.2	12
Lactating mares										
Foaling to 3 months	200		13.7	688	24	27	18	4.8	21.2	12
3 months to weaning	200		12.2	528	18	18	11	3.7	14.8	12
Working horses										
Light work[a]	200		9.3	370	13	11	8	4.3	14.1	9
Moderate work[b]	200		11.1	444	16	14	10	5.1	16.9	9
Intense work[c]	200		14.8	592	21	18	13	6.8	22.5	9
Growing horses										
Weanling, 4 months	75	0.40	7.3	365	15	16	9	1.6	5.0	3
Weanling, 6 months										
Moderate growth	95	0.30	7.6	378	16	13	7	1.8	5.7	4
Rapid growth	95	0.40	8.7	433	18	17	9	1.9	6.0	4
Yearling, 12 months										
Moderate growth	140	0.20	8.7	392	17	12	7	2.4	7.6	6
Rapid growth	140	0.30	10.3	462	19	15	8	2.5	7.9	6
Long yearling, 18 months										
Not in training	170	0.10	8.3	375	16	10	6	2.7	8.8	8
In training	170	0.10	11.6	522	22	14	8	3.7	12.2	8
Two year old, 24 months										
Not in training	185	0.05	7.9	337	13	9	5	2.8	9.4	8
In training	185	0.05	11.4	485	19	13	7	4.1	13.5	8

TABLE 5-1B   Daily Nutrient Requirements of Horses (400-kg mature weight)

Animal	Weight (kg)	Daily Gain (kg)	DE (Mcal)	Crude Protein (g)	Lysine (g)	Calcium (g)	Phos-phorus (g)	Magne-sium (g)	Potas-sium (g)	Vitamin A (10³ IU)
Mature horses										
Maintenance	400		13.4	536	19	16	11	6.0	20.0	12
Stallions	400		16.8	670	23	20	15	7.7	25.5	18
(breeding season)										
Pregnant mares										
9 months	400		14.9	654	23	28	21	7.1	23.8	24
10 months			15.1	666	23	29	22	7.3	24.2	24
11 months			16.1	708	25	31	23	7.7	25.7	24
Lactating mares										
Foaling to 3 months	400		22.9	1,141	40	45	29	8.7	36.8	24
3 months to weaning	400		19.7	839	29	29	18	6.9	26.4	24
Working horses										
Light work[a]	400		16.8	670	23	20	15	7.7	25.5	18
Moderate work[b]	400		20.1	804	28	25	17	9.2	30.6	18
Intense work[c]	400		26.8	1,072	38	33	23	12.3	40.7	18
Growing horses										
Weanling, 4 months	145	0.85	13.5	675	28	33	18	3.2	9.8	7
Weanling, 6 months										
Moderate growth	180	0.55	12.9	643	27	25	14	3.4	10.7	8
Rapid growth	180	0.70	14.5	725	30	30	16	3.6	11.1	8
Yearling, 12 months										
Moderate growth	265	0.40	15.6	700	30	23	13	4.5	14.5	12
Rapid growth	265	0.50	17.1	770	33	27	15	4.6	14.8	12
Long yearling, 18 months										
Not in training	330	0.25	15.9	716	30	21	12	5.3	17.3	15
In training	330	0.25	21.6	970	41	29	16	7.1	23.4	15
Two year old, 24 months										
Not in training	365	0.15	15.3	650	26	19	11	5.7	18.7	16
In training	365	0.15	21.5	913	37	27	15	7.9	26.2	16

TABLE 5-1C   Daily Nutrient Requirements of Horses (500-kg mature weight)

Animal	Weight (kg)	Daily Gain (kg)	DE (Mcal)	Crude Protein (g)	Lysine (g)	Calcium (g)	Phos-phorus (g)	Magne-sium (g)	Potas-sium (g)	Vitamin A (10³ IU)
Mature horses										
Maintenance	500		16.4	656	23	20	14	7.5	25.0	15
Stallions	500		20.5	820	29	25	18	9.4	31.2	22
(breeding season)										
Pregnant mares										
9 months	500		18.2	801	28	35	26	8.7	29.1	30
10 months			18.5	815	29	35	27	8.9	29.7	30
11 months			19.7	866	30	37	28	9.4	31.5	30
Lactating mares										
Foaling to 3 months	500		28.3	1,427	50	56	36	10.9	46.0	30
3 months to weaning	500		24.3	1,048	37	36	22	8.6	33.0	30
Working horses										
Light work[a]	500		20.5	820	29	25	18	9.4	31.2	22
Moderate work[b]	500		24.6	984	34	30	21	11.3	37.4	22
Intense work[c]	500		32.8	1,312	46	40	29	15.1	49.9	22
Growing horses										
Weanling, 4 months	175	0.85	14.4	720	30	34	19	3.7	11.3	8
Weanling, 6 months										
Moderate growth	215	0.65	15.0	750	32	29	16	4.0	12.7	10
Rapid growth	215	0.85	17.2	860	36	36	20	4.3	13.3	10
Yearling, 12 months										
Moderate growth	325	0.50	18.9	851	36	29	16	5.5	17.8	15
Rapid growth	325	0.65	21.3	956	40	34	19	5.7	18.2	15
Long yearling, 18 months										
Not in training	400	0.35	19.8	893	38	27	15	6.4	21.1	18
In training	400	0.35	26.5	1,195	50	36	20	8.6	28.2	18
Two year old, 24 months										
Not in training	450	0.20	18.8	800	32	24	13	7.0	23.1	20
In training	450	0.20	26.3	1,117	45	34	19	9.8	32.2	20

TABLE 5-1D Daily Nutrient Requirements of Horses (600-kg mature weight)

Animal	Weight (kg)	Daily Gain (kg)	DE (Mcal)	Crude Protein (g)	Lysine (g)	Calcium (g)	Phosphorus (g)	Magnesium (g)	Potassium (g)	Vitamin A (10³ IU)
Mature horses										
Maintenance	600		19.4	776	27	24	17	9.0	30.0	18
Stallions	600		24.3	970	34	30	21	11.2	36.9	27
(breeding season)										
Pregnant mares										
9 months	600		21.5	947	33	41	31	10.3	34.5	36
10 months			21.9	965	34	42	32	10.5	35.1	36
11 months			23.3	1,024	36	44	34	11.2	37.2	36
Lactating mares										
Foaling to 3 months	600		33.7	1,711	60	67	43	13.1	55.2	36
3 months to weaning	600		28.9	1,258	44	43	27	10.4	39.6	36
Working horses										
Light work[a]	600		24.3	970	34	30	21	11.2	36.9	27
Moderate work[b]	600		29.1	1,164	41	36	25	13.4	44.2	27
Intense work[c]	600		38.8	1,552	54	47	34	17.8	59.0	27
Growing horses										
Weanling, 4 months	200	1.00	16.5	825	35	40	22	4.3	13.0	9
Weanling, 6 months										
Moderate growth	245	0.75	17.0	850	36	34	19	4.6	14.5	11
Rapid growth	245	0.95	19.2	960	40	40	22	4.9	15.1	11
Yearling, 12 months										
Moderate growth	375	0.65	22.7	1,023	43	36	20	6.4	20.7	17
Rapid growth	375	0.80	25.1	1,127	48	41	22	6.6	21.2	17
Long yearling, 18 months										
Not in training	475	0.45	23.9	1,077	45	33	18	7.7	25.1	21
In training	475	0.45	32.0	1,429	60	44	24	10.2	33.3	21
Two year old, 24 months										
Not in training	540	0.30	23.5	998	40	31	17	8.5	27.9	24
In training	540	0.30	32.3	1,372	55	43	24	11.6	38.4	24

TABLE 5-1E Daily Nutrient Requirements of Horses (700-kg mature weight)

Animal	Weight (kg)	Daily Gain (kg)	DE (Mcal)	Crude Protein (g)	Lysine (g)	Calcium (g)	Phosphorus (g)	Magnesium (g)	Potassium (g)	Vitamin A (10³ IU)
Mature horses										
Maintenance	700		21.3	851	30	28	20	10.5	35.0	21
Stallions	700		26.6	1,064	37	32	23	12.2	40.4	32
(breeding season)										
Pregnant mares										
9 months	700		23.6	1,039	36	45	34	11.3	37.8	42
10 months			24.0	1,058	37	46	35	11.5	38.5	42
11 months			25.5	1,124	39	49	37	12.3	40.9	42
Lactating mares										
Foaling to 3 months	700		37.9	1,997	70	78	51	15.2	64.4	42
3 months to weaning	700		32.4	1,468	51	50	31	12.1	46.2	42
Working horses										
Light work[a]	700		26.6	1,064	37	32	23	12.2	40.4	32
Moderate work[b]	700		31.9	1,277	45	39	28	14.7	48.5	32
Intense work[c]	700		42.6	1,702	60	52	37	19.6	64.7	32
Growing horses										
Weanling, 4 months	225	1.10	19.7	986	41	44	25	4.8	14.6	10
Weanling, 6 months										
Moderate growth	275	0.80	20.0	1,001	42	37	20	5.1	16.2	12
Rapid growth	275	1.00	22.2	1,111	47	43	24	5.4	16.8	12
Yearling, 12 months										
Moderate growth	420	0.70	26.1	1,176	50	39	22	7.2	23.1	19
Rapid growth	420	0.85	28.5	1,281	54	44	24	7.4	23.6	19
Long yearling, 18 months										
Not in training	525	0.50	27.0	1,215	51	37	20	8.5	27.8	24
In training	525	0.50	36.0	1,615	68	49	27	11.3	36.9	24
Two year old, 24 months										
Not in training	600	0.35	26.3	1,117	45	35	19	9.4	31.1	27
In training	600	0.35	36.0	1,529	61	48	27	12.9	42.5	27

TABLE 5-1F Daily Nutrient Requirements of Horses (800-kg mature weight)

Animal	Weight (kg)	Daily Gain (kg)	DE (Mcal)	Crude Protein (g)	Lysine (g)	Calcium (g)	Phosphorus (g)	Magnesium (g)	Potassium (g)	Vitamin A (10³ IU)
Mature horses										
Maintenance	800		22.9	914	32	32	22	12.0	40.0	24
Stallions	800		28.6	1,143	40	35	25	13.1	43.4	36
(breeding season)										
Pregnant mares										
9 months	800		25.4	1,116	39	48	37	12.2	40.6	48
10 months			25.8	1,137	40	49	37	12.4	41.3	48
11 months			27.4	1,207	42	52	40	13.2	43.9	48
Lactating mares										
Foaling to 3 months	800		41.9	2,282	81	90	58	17.4	73.6	48
3 months to weaning	800		35.5	1,678	60	58	36	13.8	52.8	48
Working horses										
Light work[a]	800		28.6	1,143	40	35	25	13.1	43.4	36
Moderate work[b]	800		34.3	1,372	48	42	30	15.8	52.1	36
Intense work[c]	800		45.7	1,829	64	56	40	21.0	69.5	36
Growing horses										
Weanling, 4 months	250	1.20	21.4	1,070	45	48	27	5.3	16.1	11
Weanling, 6 months										
Moderate growth	305	0.90	22.0	1,100	46	41	23	5.7	18.0	14
Rapid growth	305	1.10	24.2	1,210	51	47	26	6.0	18.6	14
Yearling, 12 months										
Moderate growth	460	0.80	28.7	1,291	55	44	24	7.9	25.4	21
Rapid growth	460	0.95	31.0	1,396	59	49	27	8.1	25.9	21
Long yearling, 18 months										
Not in training	590	0.60	30.2	1,361	57	43	24	9.6	31.3	27
In training	590	0.60	39.8	1,793	76	56	31	12.6	41.2	27
Two year old, 24 months										
Not in training	675	0.40	28.7	1,220	49	40	22	10.6	35.0	30
In training	675	0.40	39.1	1,662	66	54	30	14.5	47.6	30

TABLE 5-1G Daily Nutrient Requirements of Horses (900-kg mature weight)

Animal	Weight (kg)	Daily Gain (kg)	DE (Mcal)	Crude Protein (g)	Lysine (g)	Calcium (g)	Phosphorus (g)	Magnesium (g)	Potassium (g)	Vitamin A (10³ IU)
Mature horses										
Maintenance	900		24.1	966	34	36	25	13.5	45.0	27
Stallions	900		30.2	1,207	42	37	26	13.9	45.9	40
(breeding season)										
Pregnant mares										
9 months	900		26.8	1,179	41	51	39	12.9	42.9	54
10 months			27.3	1,200	42	52	39	13.1	43.6	54
11 months			29.0	1,275	45	55	42	13.9	46.3	54
Lactating mares										
Foaling to 3 months	900		45.5	2,567	89	101	65	19.6	82.8	54
3 months to weaning	900		38.4	1,887	66	65	40	15.5	59.4	54
Working horses										
Light work[a]	900		30.2	1,207	42	37	26	13.9	45.9	40
Moderate work[b]	900		36.2	1,448	51	44	32	16.7	55.0	40
Intense work[c]	900		48.3	1,931	68	59	42	22.2	73.4	40
Growing horses										
Weanling, 4 months	275	1.30	23.1	1,154	48	53	29	5.8	17.7	12
Weanling, 6 months										
Moderate growth	335	0.95	23.4	1,171	49	44	24	6.2	19.6	15
Rapid growth	335	1.15	25.6	1,281	54	50	28	6.5	20.2	15
Yearling, 12 months										
Moderate growth	500	0.90	31.2	1,404	59	49	27	8.6	27.7	22
Rapid growth	500	1.05	33.5	1,509	64	54	30	8.8	28.2	22
Long yearling, 18 months										
Not in training	665	0.70	33.6	1,510	64	49	27	10.9	35.4	30
In training	665	0.70	43.9	1,975	83	64	35	14.2	46.2	30
Two year old, 24 months										
Not in training	760	0.45	31.1	1,322	53	45	25	12.0	39.4	34
In training	760	0.45	42.2	1,795	72	61	34	16.2	53.4	34

NOTE: Mares should gain weight during late gestation to compensate for tissue deposition. However, nutrient requirements are based on maintenance body weight.

[a] Examples are horses used in Western and English pleasure, bridle path hack, equitation, etc.

[b] Examples are horses used in ranch work, roping, cutting, barrel racing, jumping, etc.

[c] Examples are horses in race training, polo, etc.

TABLE 5-2A   Nutrient Concentrations in Total Diets for Horses and Ponies (dry matter basis)

	Digestible Energy[a]		Diet Proportions		Crude Protein (%)	Lysine (%)	Cal-cium (%)	Phos-phorus (%)	Mag-nesium (%)	Potas-sium (%)	Vitamin A	
	(Mcal/kg)	(Mcal/lb)	Conc. (%)	Hay (%)							(IU/kg)	(IU/lb)
Mature horses												
Maintenance	2.00	0.90	0	100	8.0	0.28	0.24	0.17	0.09	0.30	1830	830
Stallions	2.40	1.10	30	70	9.6	0.34	0.29	0.21	0.11	0.36	2640	1200
Pregnant mares												
9 months	2.25	1.00	20	80	10.0	0.35	0.43	0.32	0.10	0.35	3710	1680
10 months	2.25	1.00	20	80	10.0	0.35	0.43	0.32	0.10	0.36	3650	1660
11 months	2.40	1.10	30	70	10.6	0.37	0.45	0.34	0.11	0.38	3650	1660
Lactating mares												
Foaling to 3 months	2.60	1.20	50	50	13.2	0.46	0.52	0.34	0.10	0.42	2750	1250
3 months to weaning	2.45	1.15	35	65	11.0	0.37	0.36	0.22	0.09	0.33	3020	1370
Working horses												
Light work[b]	2.45	1.15	35	65	9.8	0.35	0.30	0.22	0.11	0.37	2690	1220
Moderate work[c]	2.65	1.20	50	50	10.4	0.37	0.31	0.23	0.12	0.39	2420	1100
Intense work[d]	2.85	1.30	65	35	11.4	0.40	0.35	0.25	0.13	0.43	1950	890
Growing horses												
Weanling, 4 months	2.90	1.40	70	30	14.5	0.60	0.68	0.38	0.08	0.30	1580	720
Weanling, 6 months												
Moderate growth	2.90	1.40	70	30	14.5	0.61	0.56	0.31	0.08	0.30	1870	850
Rapid growth	2.90	1.40	70	30	14.5	0.61	0.61	0.34	0.08	0.30	1630	740
Yearling, 12 months												
Moderate growth	2.80	1.30	60	40	12.6	0.53	0.43	0.24	0.08	0.30	2160	980
Rapid growth	2.80	1.30	60	40	12.6	0.53	0.45	0.25	0.08	0.30	1920	870
Long yearling, 18 months												
Not in training	2.50	1.15	45	55	11.3	0.48	0.34	0.19	0.08	0.30	2270	1030
In training	2.65	1.20	50	50	12.0	0.50	0.36	0.20	0.09	0.30	1800	820
Two year old, 24 months												
Not in training	2.45	1.15	35	65	10.4	0.42	0.31	0.17	0.09	0.30	2640	1200
In training	2.65	1.20	50	50	11.3	0.45	0.34	0.20	0.10	0.32	2040	930

[a] Values assume a concentrate feed containing 3.3 Mcal/kg and hay containing 2.00 Mcal/kg of dry matter.

[b] Examples are horses used in Western and English pleasure, bridle path hack, equitation, etc.

[c] Examples are horses used in ranch work, roping, cutting, barrel racing, jumping, etc.

[d] Examples are race training, polo, etc.

TABLE 5-2B   Nutrient Concentrations in Total Diets for Horses and Ponies (90% dry matter basis)

	Digestible Energy[a]		Diet Proportions		Crude Protein (%)	Lysine (%)	Cal-cium (%)	Phos-phorus (%)	Mag-nesium (%)	Potas-sium (%)	Vitamin A	
	(Mcal/kg)	(Mcal/lb)	Conc. (%)	Hay (%)							(IU/kg)	(IU/lb)
Mature horses												
Maintenance	1.80	0.80	0	100	7.2	0.25	0.21	0.15	0.08	0.27	1650	750
Stallions	2.15	1.00	30	70	8.6	0.30	0.26	0.19	0.10	0.33	2370	1080
Pregnant mares												
9 months	2.00	0.90	20	80	8.9	0.31	0.39	0.29	0.10	0.32	3330	1510
10 months	2.00	0.90	20	80	9.0	0.32	0.39	0.30	0.10	0.33	3280	1490
11 months	2.15	1.00	30	70	9.5	0.33	0.41	0.31	0.10	0.35	3280	1490
Lactating mares												
Foaling to 3 months	2.35	1.10	50	50	12.0	0.41	0.47	0.30	0.09	0.38	2480	1130
3 months to weaning	2.20	1.05	35	65	10.0	0.34	0.33	0.20	0.08	0.30	2720	1240
Working horses												
Light work[b]	2.20	1.05	35	65	8.8	0.32	0.27	0.19	0.10	0.34	2420	1100
Moderate work[c]	2.40	1.10	50	50	9.4	0.35	0.28	0.22	0.11	0.36	2140	970
Intense work[d]	2.55	1.20	65	35	10.3	0.36	0.31	0.23	0.12	0.39	1760	800
Growing horses												
Weanling, 4 months	2.60	1.25	70	30	13.1	0.54	0.62	0.34	0.07	0.27	1420	650
Weanling, 6 months												
Moderate growth	2.60	1.25	70	30	13.0	0.55	0.50	0.28	0.07	0.27	1680	760
Rapid growth	2.60	1.25	70	30	13.1	0.55	0.55	0.30	0.07	0.27	1470	670
Yearling, 12 months												
Moderate growth	2.50	1.15	60	40	11.3	0.48	0.39	0.21	0.07	0.27	1950	890
Rapid growth	2.50	1.15	60	40	11.3	0.48	0.40	0.22	0.07	0.27	1730	790
Long yearling, 18 months												
Not in training	2.30	1.05	45	55	10.1	0.43	0.31	0.17	0.07	0.27	2050	930
In training	2.40	1.10	50	50	10.8	0.45	0.32	0.18	0.08	0.27	1620	740
Two year old, 24 months												
Not in training	2.20	1.00	35	65	9.4	0.38	0.28	0.15	0.08	0.27	2380	1080
In training	2.40	1.10	50	50	10.1	0.41	0.31	0.17	0.09	0.29	1840	840

[a] Values assume a concentrate feed containing 3.3 Mcal/kg and hay containing 2.00 Mcal/kg of dry matter.

[b] Examples are horses used in Western and English pleasure, bridle path hack, equitation, etc.

[c] Examples are horses used in ranch work, roping, cutting, barrel racing, jumping, etc.

[d] Examples are race training, polo, etc.

TABLE 5-3   Other Minerals and Vitamins for Horses and Ponies (on a dry matter basis)

	Adequate Concentrations in Total Rations				
	Main-tenance	Pregnant and Lactating Mares	Growing Horses	Working Horses	Maximum Tolerance Levels
Minerals					
Sodium (%)	0.10	0.10	0.10	0.30	3[a]
Sulfur (%)	0.15	0.15	0.15	0.15	1.25
Iron (mg/kg)	40	50	50	40	1,000
Manganese (mg/kg)	40	40	40	40	1,000
Copper (mg/kg)	10	10	10	10	800
Zinc (mg/kg)	40	40	40	40	500
Selenium (mg/kg)	0.1	0.1	0.1	0.1	2.0
Iodine (mg/kg)	0.1	0.1	0.1	0.1	5.0
Cobalt (mg/kg)	0.1	0.1	0.1	0.1	10
Vitamins					
Vitamin A (IU/kg)	2,000	3,000	2,000	2,000	16,000
Vitamin D (IU/kg)[b]	300	600	800	300	2,200
Vitamin E (IU/kg)	50	80	80	80	1,000
Vitamin K (mg/kg)	[c]				
Thiamin (mg/kg)	3	3	3	5	3,000
Riboflavin (mg/kg)	2	2	2	2	
Niacin (mg/kg)					
Pantothenic acid (mg/kg)					
Pyridoxine (mg/kg)					
Biotin (mg/kg)					
Folacin (mg/kg)					
Vitamin B_{12} (μg/kg)					
Ascorbic acid (mg/kg)					
Choline (mg/kg)					

[a] As sodium chloride.
[b] Recommendations for horses not exposed to sunlight or to artificial light with an emission spectrum of 280–315 nm.
[c] Blank space indicates that data are insufficient to determine a requirement or maximum tolerable level.

TABLE 5-4   Expected Feed Consumption by Horses (% body weight)[a]

	Forage	Concentrate	Total
Mature horses			
Maintenance	1.5–2.0	0–0.5	1.5–2.0
Mares, late gestation	1.0–1.5	0.5–1.0	1.5–2.0
Mares, early lactation	1.0–2.0	1.0–2.0	2.0–3.0
Mares, late lactation	1.0–2.0	0.5–1.5	2.0–2.5
Working horses			
Light work	1.0–2.0	0.5–1.0	1.5–2.5
Moderate work	1.0–2.0	0.75–1.5	1.75–2.5
Intense work	0.75–1.5	1.0–2.0	2.0–3.0
Young horses			
Nursing foal, 3 months	0	1.0–2.0	2.5–3.5
Weanling foal, 6 months	0.5–1.0	1.5–3.0	2.0–3.5
Yearling foal, 12 months	1.0–1.5	1.0–2.0	2.0–3.0
Long yearling, 18 months	1.0–1.5	1.0–1.5	2.0–2.5
Two year old (24 months)	1.0–1.5	1.0–1.5	1.75–2.5

[a] Air-dry feed (about 90% DM).

6 Feed Composition Tables

TABLE 6-1A   Composition of Feeds (Excluding Vitamins) Commonly Used in Horse Diets

Entry No.	Feed Name	International Feed Number[a]	Dry Matter (%)	DE[b] (Mcal/kg)	(Mcal/lb)	Crude Protein (%)	Lysine (%)	Ether Extract (%)	Fiber (%)	NDF (%)	ADF (%)	Ash (%)
	ALFALFA *Medicago sativa*											
001	—fresh, late vegetative	2-00-181	23.2	0.68	0.31	5.1	0.29	0.7	5.6	7.2	5.6	2.4
002			100.0	2.94	1.34	22.2	1.24	2.9	24.2	30.9	24.0	10.2
		N	14			17	1	4	4	14	12	6
		SD	3.39			2.00		0.947	2.29	4.79	3.66	0.826
003	—fresh, full bloom	2-00-188	23.8	0.55	0.25	4.6		0.6	7.2	9.2	8.6	2.6
004			100.0	2.32	1.05	19.3		2.8	30.4	38.6	35.9	10.9
		N	8			8		2	2	12	2	8
		SD	3.88			3.70		0.565	1.83	6.14	2.82	2.35
005	—hay, sun-cured, early bloom	1-00-059	90.5	2.24	1.02	18.0	0.81	2.6	20.8	35.6	28.9	8.4
006			100.0	2.48	1.13	19.9	0.90	2.9	23.0	39.3	31.9	9.2
		N	43			63	1	28	29	14	15	36
		SD	1.92			2.25		1.35	3.98	3.58	2.40	1.61
007	—hay, sun-cured, midbloom	1-00-063	91.0	2.07	0.94	17.0		2.4	25.5	42.9	33.4	7.8
008			100.0	2.28	1.03	18.7		2.6	28.0	47.1	36.7	8.5
		N	60			56		23	22	22	26	41
		SD	1.88			2.93		1.82	4.25	6.53	2.58	1.48
009	—hay, sun-cured, full bloom	1-00-068	90.9	1.97	0.89	15.5	0.79	1.8	27.3	44.3	35.2	7.1
010			100.0	2.17	0.98	17.0	0.87	2.0	30.1	48.8	38.7	7.8
		N	21			20	1	12	14	10	9	16
		SD	2.06			2.50		1.73	4.27	3.49	2.42	1.07
011	—meal, dehydrated, 15% protein	1-00-022	90.4	2.00	0.91	15.6	0.63	2.2	26.2	50.1	33.9	8.9
012			100.0	2.21	1.00	17.3	0.69	2.4	29.0	55.4	37.5	9.9
		N	23			21	4	13	18	1	2	12
		SD	2.18			1.75	0.077	0.438	3.17		1.47	0.934
013	—meal, dehydrated, 17% protein	1-00-023	91.8	2.16	0.98	17.4	0.85	2.8	24.0	41.3	31.5	9.8
014			100.0	2.36	1.07	18.9	0.92	3.0	26.2	45.0	34.3	10.6
		N	72			50	24	37	46	1	2	21
		SD	1.50			0.677	0.169	0.493	2.25		0.951	0.613
	ALYCECLOVER *Alysicarpus vaginalis*											
015	—hay, sun-cured	1-00-361	89.7	1.64	0.75	10.9		1.6	36.2			5.7
016			100.0	1.83	0.83	12.2		1.7	40.3			6.4
		N	4			3		2	3			2
		SD	0.901			0.464		0.068	1.30			0.147
	BAHIAGRASS *Paspalum notatum*											
017	—fresh	2-00-464	28.7	0.61	0.26	3.6		0.5	8.7	21.0	10.2	3.2
018			100.0	2.03	0.92	12.6		1.6	30.4	73.1	35.5	11.1
		N	9			6		2	3	3	3	2
		SD	1.91			4.14		0.047	1.83	0.233	0.493	1.27
019	—hay, sun-cured	1-00-462	90.0	1.75	0.79	8.5		1.8	28.1	66.6	33.9	5.7
020			100.0	1.94	0.88	9.5		2.0	31.2	73.9	37.7	6.3
		N	4			6		2	6	3	3	1
		SD	0.420			2.30		0.141	2.33	1.67	2.86	
021	—hay, sun-cured, late vegetative	1-20-787	91.0	1.70	0.77	8.9		2.1	29.7	66.4	37.9	7.1
022			100.0	1.87	0.85	9.8		2.3	32.7	73.0	41.6	7.8
		N	1			2		2	2	1	2	2
		SD				0.424		0.848	0.495		5.09	2.54
023	—hay, sun-cured, early bloom	1-06-138	91.0	1.61	0.73	6.4		1.4	30.9	69.2	38.2	8.5
024			100.0	1.77	0.80	7.0		1.5	34.0	76.0	42.0	9.3
		N	1			1		1	1	1	1	1
		SD										
	BARLEY *Hordeum vulgare*											
025	—grain	4-00-549	88.6	3.26	1.49	11.7	0.40	1.8	4.9	16.8	6.2	2.4
026			100.0	3.68	1.67	13.2	0.45	2.0	5.6	19.0	7.0	2.7
		N	237			304	67	208	210	25	13	217
		SD	2.08			1.90	0.099	0.465	1.09	3.33	5.22	0.335
027	—grain, Pacific coast	4-07-939	88.6	3.17	1.48	9.7	0.27	2.0	6.0	18.6	8.0	2.4
028			100.0	3.58	1.67	11.0	0.30	2.2	6.8	21.0	9.0	2.7
		N	17			14	10	9	11	3	4	8
		SD	0.773			0.640	0.052	0.135	1.25	0.503	0.997	0.139
029	—hay, sun-cured	1-00-495	88.4	1.78	0.81	7.8		1.9	23.6			6.6
030			100.0	2.01	0.92	8.8		2.1	26.7			7.5
		N	10			6		6	5			4
		SD	2.17			0.426		0.120	0.694			0.206
031	—straw	1-00-498	91.2	1.47	0.67	4.0		1.7	37.8	66.1	44.5	6.8
032			100.0	1.62	0.73	4.4		1.9	41.5	72.5	48.8	7.5
		N	29			35		7	26	2	3	8
		SD	3.31			0.907		0.269	4.03	1.83	4.65	1.40

Entry No.	Macrominerals (%)						Microminerals (mg/kg)						
	Calcium	Phosphorus	Magnesium	Potassium	Sodium	Sulfur	Copper	Iodine	Iron	Manganese	Selenium	Zinc	Cobalt
001	0.40	0.07	0.08	0.53	0.05	0.08	2.5		26	9			0.04
002	1.71	0.30	0.36	2.27	0.21	0.36	10.8		111	41			0.17
	10	10	10	10	2	9	1		1	2			1
	0.488	0.038	0.105	0.507	0.015	0.097				18			
003	0.28	0.06	0.10	0.86	0.04	0.07	3.6		70	10		8	0.12
004	1.19	0.26	0.40	3.62	0.16	0.31	14.9		293	41		32	0.49
	6	6	6	6	6	1	5		6	6		6	5
	0.242	0.036	0.106	0.892	0.069		2.3		232	35.2		16.2	0.057
005	1.28	0.19	0.31	2.32	0.14	0.27	11.4	0.15	205	33	0.50	27	0.26
006	1.41	0.21	0.34	2.56	0.15	0.30	12.7	0.17	227	36	0.55	30	0.29
	98	91	93	96	7	1	93	1	97	95	86	97	9
	0.394	0.047	0.100	0.617	0.135		3.0		137	25.5	0.395	7.6	0.238
007	1.24	0.22	0.32	1.42	0.11	0.26	16.1	0.15	204	25		28	0.36
008	1.37	0.24	0.35	1.56	0.12	0.28	17.7	0.16	225	28		31	0.39
	9	13	7	8	5	3	3	1	4	4		3	2
	0.287	0.047	0.111	0.510	0.053	0.028	5.6		182	7.7		14.1	0.047
009	1.08	0.22	0.25	1.42	0.06	0.25	9.0	0.12	141	38		24	0.21
010	1.19	0.24	0.27	1.56	0.07	0.27	9.9	0.13	155	42		26	0.23
	6	7	6	7	3	1	6	1	8	6		4	4
	0.144	0.078	0.112	0.758	0.073		4.2		28.1	8.6		2.8	0.280
011	1.25	0.23	0.26	2.22	0.07	0.19	9.5	0.12	280	28	0.28	19	0.17
012	1.38	0.25	0.29	2.46	0.08	0.21	10.5	0.13	309	31	0.31	21	0.19
	5	5	5	6	4	4	2	1	3	2	2	2	1
	0.073	0.026	0.047	0.143	0.006	0.019	1.7		54.7	2.5	0.320	1.4	
013	1.38	0.23	0.29	2.40	0.10	0.22	8.6	0.15	405	31	0.33	19	0.30
014	1.51	0.25	0.32	2.61	0.11	0.24	9.3	0.16	441	34	0.36	21	0.33
	25	28	12	11	10	8	6	1	7	7	2	5	3
	0.132	0.024	0.041	0.292	0.050	0.029	1.7		76.4	3.7	0.396	7.5	0.041
015													
016													
017	0.13	0.09	0.08	0.44			2.1		24	21	0.02	8	
018	0.44	0.30	0.27	1.53			7.2		85	75	0.06	29	
	3	3	2	2			2		1	1		1	
	0.040	0.173	0.028	0.106			1.8						
019	0.45	0.20	0.17						54.01				
020	0.50	0.22	0.19						60.00				
	1	1	1						1				
021	0.25	0.19	0.25	1.64									
022	0.28	0.21	0.27	1.80									
	1	1	1	1									
023	0.24	0.18	0.23	1.46									
024	0.26	0.20	0.25	1.60									
	1	1	1	1									
025	0.05	0.34	0.13	0.44	0.03	0.15	8.2	0.04	73.53	16	0.18	17	0.17
026	0.05	0.38	0.15	0.50	0.03	0.17	9.2	0.05	83.02	18	0.20	19	0.19
	117	134	73	71	19	14	45	1	59	60	54	65	6
	0.018	0.073	0.018	0.132	0.017	0.009	4.1		30.2	3.7	0.152	18.0	0.229
027	0.05	0.34	0.12	0.51	0.02	0.14	8.1		86.14	16	0.10	15	0.09
028	0.05	0.38	0.14	0.58	0.02	0.16	9.1		97.19	18	0.11	17	0.10
	12	12	3	5	4	3	2		3	3	1	3	1
	0.013	0.045	0.003	0.056	0.001	0.012	0.65		32.5	0.4		0.27	
029	0.21	0.25	0.14	1.30	0.12	0.15	3.9		265.31	35			0.06
030	0.24	0.28	0.16	1.47	0.14	0.17	4.4		299.96	39			0.07
	5	5	2	3	1	2	1		1	3			1
	0.039	0.028	0.042	0.025		0.002				0.3			
031	0.27	0.06	0.21	2.16	0.13	0.16	4.9		183.09	15		7	0.06
032	0.30	0.07	0.23	2.36	0.14	0.17	5.40		200.78	16		7	0.07
	34	40	22	22	5	5	18		20	4		17	1
	0.088	0.032	0.052	0.480	0.003	0.012	1.3		72.0	0.7		0.583	

Continues

TABLE 6-1A   Composition of Feeds (Excluding Vitamins) Commonly Used in Horse Diets—*Continued*

Entry No.	Feed Name	International Feed Number[a]	Dry Matter (%)	DE[b] (Mcal/kg)	DE[b] (Mcal/lb)	Crude Protein (%)	Lysine (%)	Ether Extract (%)	Fiber (%)	NDF (%)	ADF (%)	Ash (%)
	BEET, SUGAR *Beta vulgaris altissima*											
	—pulp, dehydrated											
033		4-00-669	91.0	2.33	1.06	8.9	0.54	0.5	18.2	40.5	25.0	4.9
034			100.0	2.56	1.16	9.8	0.60	0.6	20.0	44.6	27.5	5.3
		N	47			31	6	25	29	2	5	22
		SD	1.37			1.04	0.151	0.151	2.40	20.4	6.79	1.29
	BERMUDAGRASS, COASTAL *Cynodon dactylon*											
	—fresh											
035		2-00-719	30.3	0.72	0.33	3.8		1.1	8.6	22.2	11.1	2.4
036			100.0	2.38	1.08	12.6		3.7	28.4	73.3	36.8	8.1
		N	15			48		10	11	41	41	34
		SD	6.91			2.88		0.954	1.77	5.10	4.64	1.86
037	—hay, sun-cured, 15–28 days' growth	1-09-207	88.4	1.92	0.87	10.6	0.38	2.4	26.7	64.5	30.0	6.7
038			100.0	2.17	0.98	12.0	0.45	2.7	30.2	73.0	34.0	7.5
		N	4			17		16	16	6	6	16
		SD	0.326			1.84		0.380	2.30	2.54	1.84	0.860
039	—hay, sun-cured, 29–42 days' growth	1-09-209	93.0	1.96	0.89	10.9		2.4	28.0	68.0	32.7	6.2
040			100.0	2.10	0.95	12.0		2.6	31.0	75.0	36.2	6.9
		N	4			22		18	18	5	5	18
		SD	1.61			3.05		0.68	2.59	1.60	1.40	1.18
041	—hay, sun-cured, 43–56 days' growth	1-09-210	93.0	1.74	0.79	7.3	0.28	2.5	30.4	71.3	35.7	7.5
042			100.0	1.87	0.85	7.8	0.30	2.7	32.6	76.6	38.3	8.0
		N	1			4		2	2	3	3	2
		SD				1.19		1.83	4.73	2.45	4.18	1.34
	BLUEGRASS, KENTUCKY *Poa pratensis*											
043	—fresh, early vegetative	2-00-777	30.8	0.64	0.29	5.4		1.1	7.8			2.9
044			100.0	2.09	0.95	17.4		3.5	25.2			9.4
		N	4			2		2	2			1
		SD	0.694			0.144		0.070	0.211			
045	—fresh, milk stage	2-00-782	42.0	0.75	0.34	4.9		1.5	12.7			3.1
046			100.0	1.78	0.81	11.6		3.6	30.3			7.3
		N	1			1		1	1			1
		SD										
047	—hay, sun-cured, full bloom	1-00-772	92.1	1.58	0.72	8.2		3.0	29.9			5.4
048			100.0	1.71	0.78	8.9		3.3	32.5			5.9
		N	2			1		1	1			1
		SD										
	BREWERS GRAINS, DEHYDRATED											
049		5-02-141	92.0	2.53	1.15	23.4	0.88	5.9	13.7	42.3	22.1	4.4
050			100.0	2.75	1.25	25.4	0.96	6.5	14.9	46.0	24.0	4.8
	BROME, SMOOTH *Bromus inermis*											
051	—fresh, early vegetative	2-00-956	26.1	0.66	0.31	5.6		1.0	6.0	12.5	8.1	2.7
052			100.0	2.59	1.17	21.3		4.0	23.0	47.9	31.0	10.4
		N	8			6		3	3	5	5	6
		SD	6.39			2.47		0.351	0.530	3.63	3.16	0.459
053	—fresh, mature	2-08-364	54.9	0.89	0.40	3.4		1.3	19.1	35.4		3.8
054			100.0	1.62	0.73	6.1		2.4	34.8	64.5		6.9
		N	5			4		3	3	1		3
		SD	1.69			0.577		0.078	1.39			0.120
055	—hay, sun-cured, midbloom	1-05-633	87.6	1.87	0.85	12.6		1.9	28.0	50.5	32.2	9.5
056			100.0	2.13	0.97	14.4		2.2	31.9	57.7	36.8	10.9
		N	2			4		3	3	1	3	3
		SD				3.22		0.157	3.21		4.58	1.75
057	—hay, sun-cured, mature	1-00-944	92.6	1.57	0.71	5.6		1.8	29.8	65.3	41.5	6.7
058			100.0	1.69	0.77	6.0		2.0	32.2	70.5	44.8	7.2
		N	6			2		1	1	1	1	2
		SD	0.539			0.283			2.82			1.41
	CANARYGRASS, REED *Phalaris arundinacea*											
059	—fresh	2-01-113	22.8	0.58	0.26	3.9		0.9	5.6	10.6	6.5	2.3
060			100.0	2.54	1.15	17.0		4.1	24.4	46.4	28.3	10.2
		N	4			3		2	2	1	1	3
		SD	4.89			3.65		0.494	3.39			1.85
061	—hay, sun-cured	1-01-104	89.3	1.78	0.81	9.1		2.7	30.2	62.9	32.7	7.3
062			100.0	2.00	0.91	10.2		3.0	33.9	70.5	36.6	8.1
		N	10			14		10	10	6	6	10
		SD	2.08			2.06		0.639	3.80	1.14	0.783	0.804

Entry No.	Macrominerals (%)						Microminerals (mg/kg)						
	Calcium	Phosphorus	Magnesium	Potassium	Sodium	Sulfur	Copper	Iodine	Iron	Manganese	Selenium	Zinc	Cobalt
033	0.62	0.09	0.26	0.20	0.18	0.20	12.5		266.69	34	0.11	1.0	0.07
034	0.68	0.10	0.28	0.22	0.20	0.22	13.8		293.00	38	0.12	1.0	0.08
	18	23	21	12	8	9	5		13	10	1	3	3
	0.069	0.012	0.047	0.073	0.073	0.010	0.07		62.8	1.3		0.03	0.041
035	0.15	0.08											
036	0.49	0.27											
	8	8											
	0.072	0.033											
037	0.35	0.24	0.18	1.90									
038	0.40	0.27	0.21	2.20									
	1	1	1	1									
039	0.30	0.19	0.11	1.58									
040	0.32	0.20	0.16	1.70									
	1	1	1	1									
041	0.24	0.17	0.12	1.21									
042	0.26	0.18	0.13	1.30									
	1	1	1	1									
043	0.15	0.14	0.05	0.70	0.04	0.05			92				
044	0.50	0.44	0.18	2.27	0.14	0.17			300				
	2	2	2	2	1	1			1				
	0.091	0.042	0.007	0.014									
045													
046													
047	0.24	0.25		1.40									
048	0.26	0.27		1.52									
	1	1		1									
049	0.30	0.50	0.15	0.08	0.21	0.29	21.2	0.06	245	37	0.70	28	0.07
050	0.33	0.55	0.16	0.09	0.23	0.32	23.0	0.07	266	40	0.76	30	0.08
051	0.14	0.12	0.08	0.82		0.05						6	
052	0.55	0.45	0.32	3.16		0.20						21	
	2	2	1	1		1						1	
	0.098	0.176											
053	0.14	0.09					1.2						
054	0.26	0.16					2.2						
	2	3					1						
	0.009	0.036											
055	0.25	0.25	0.09	1.74	0.01		21.9		80	35		26	0.51
056	0.29	0.28	0.10	1.99	0.01		25.0		91	40		30	0.58
	1	1	1	1	1		1		1	1		1	1
057	0.24	0.20	0.11	1.71	0.01		9.6		74	68		22	0.17
058	0.26	0.22	0.12	1.85	0.01		10.4		80	73		24	0.19
	3	2	3	3	2		2		2	2		1	2
	0.149	0.007	0.066	0.803			5.1		28.2	45.8			0.061
059	0.08	0.08		0.83									
060	0.36	0.33		3.64									
	2	2		1									
	0.056	0.042											
061	0.32	0.21	0.19	2.60	0.01	0.12	10.6		134	82		16	
062	0.36	0.24	0.22	2.91	0.02	0.14	11.9		150	92		18	
	12	12	8	8	2	1	1		1	1		1	
	0.095	0.040	0.062	0.471	0.007								

Continues

TABLE 6-1A   Composition of Feeds (Excluding Vitamins) Commonly Used in Horse Diets—*Continued*

Entry No.	Feed Name	International Feed Number[a]	Dry Matter (%)	DE[b] (Mcal/kg)	(Mcal/lb)	Crude Protein (%)	Lysine (%)	Ether Extract (%)	Fiber (%)	NDF (%)	ADF (%)	Ash (%)
	CANOLA *Brassica napus-Brassica campestris*											
063	—seeds, meal solvent-	5-06-146	90.8	2.83	1.28	37.1	2.08	2.8	11.0	21.8	15.8	6.4
064	extracted		100.0	3.11	1.41	40.9	2.29	3.0	12.1	24.0	17.4	7.1
		N	13			33	25	28	19	16	2	26
		SD	2.41			2.02	0.169	1.04	2.16	1.46	1.62	0.416
	CARROT *Daucus* spp.											
065	—roots, fresh	4-01-145	11.5	0.43	0.20	1.2		0.2	1.1	1.4	1.3	1.0
066			100.0	3.78	1.72	10.0		1.3	9.5	12	11	8.4
		N	9			5		6	6	1	1	5
		SD	1.21			0.842		0.501	1.32			1.95
	CITRUS *Citrus* spp.											
067	—pomace without	4-01-237	91.1	2.56	1.16	6.1	0.20	3.4	11.6	21.0	21.0	6.0
068	fines, dehydrated		100.0	2.81	1.27	6.7	0.22	3.7	12.8	23.0	23.0	6.6
		N	275			365	4	260	314	1	1	335
		SD	1.52			0.398	0.000	0.862	1.19			0.808
	CLOVER, ALSIKE *Trifolium hybridum*											
069	—fresh, early	2-01-314	18.9	0.47	0.21	4.5		0.6	3.3			2.4
070	vegetative		100.0	2.49	1.13	24.1		3.2	17.5			12.8
		N	3			1		1	1			1
		SD	0.116									
071	—hay, sun-cured	1-01-313	87.7	1.71	0.78	12.4		2.4	26.2			7.6
072			100.0	1.95	0.89	14.2		2.8	29.9			8.7
		N	9			2		3	3			3
		SD	1.32			0.770		0.381	0.488			0.087
	CLOVER, LADINO *Trifolium repens*											
073	—fresh, early	2-01-380	19.3	0.48	0.22	5.0		0.9	2.7			2.3
074	vegetative		100.0	2.50	1.14	25.8		4.6	13.9			11.9
		N	4			3		3	3			3
		SD	1.44			1.21		1.87	0.404			1.38
075	—hay, sun-cured	1-01-378	89.1	1.96	0.89	20.0		2.4	18.5	32.1	28.5	8.4
076			100.0	2.20	1.00	22.4		2.7	20.8	36.0	32.0	9.4
		N	5			4		3	3	1	1	2
		SD	2.71			1.18		0.750	2.90			0.161
	CLOVER, RED *Trifolium pratense*											
077	—fresh, early bloom	2-01-428	19.6	0.50	0.22	4.1		1.0	4.6	7.8	6.1	2.0
078			100.0	2.53	1.15	20.8		5.0	23.2	40.0	31.0	10.2
		N	5			3		2	3	1	1	2
		SD	0.464			3.06		0.070	4.25			0.567
079	—fresh, full bloom	2-01-429	26.2	0.66	0.27	3.8		0.8	6.8	11.3	9.2	2.0
080			100.0	2.25	1.02	14.6		2.9	26.1	43.0	35.0	7.8
		N	4			3		2	2	1	1	2
		SD	3.00			0.461		1.55	5.02			0.705
081	—fresh, regrowth,	2-28-255	24.0	0.77	0.35	5.4				6.4	4.8	
082	early vegetative		100.0	3.19	1.45	22.3				26.7	19.9	
		N	8			8				8	8	
		SD	2.58			2.12				2.80	2.91	
083	—hay, sun-cured	1-01-415	88.4	1.96	0.89	13.2		2.5	27.1	41.4	31.8	6.7
084			100.0	2.22	1.01	15.0		2.8	30.7	46.9	36.0	7.5
		N	21			13		11	11	2	2	9
		SD	1.91			1.91		0.322	3.96	12.9	9.19	0.889
	CORN, DENT YELLOW *Zea mays indentata*											
085	—aerial part with ears,	1-28-231	81.0	1.67	0.76	7.2		1.9	20.4	38.9	20.2	5.5
086	sun-cured (fodder)		100.0	2.06	0.94	8.9		2.4	25.2	48.0	25.0	6.8
		N	1			1		1	1	1	1	1
		SD										
087	—aerial part without	1-28-233	87.7	1.42	0.65	4.1		1.1	30.3	69.5	44.0	6.2
088	ears, without husks,		100.0	1.62	0.74	4.6		1.3	34.6	79.3	50.1	7.0
	sun-cured (stover,	N	4			3		1	1	3	3	2
	straw)	SD	2.43			1.38				8.34	6.65	0.374
089	—cobs, ground	1-28-234	90.1	1.22	0.56	2.5		0.6	31.9	78.4	35.6	1.6
090			100.0	1.36	0.62	2.8		0.6	35.4	87.0	39.5	1.8
		N	3			3		3	3	2	2	1
		SD	0.246			0.278		0.148	0.405	2.82	6.36	

Entry No.	Macrominerals (%)						Microminerals (mg/kg)						
	Calcium	Phosphorus	Magnesium	Potassium	Sodium	Sulfur	Copper	Iodine	Iron	Manganese	Selenium	Zinc	Cobalt
063	0.63	1.18	0.55	1.22	0.01	1.23	77		85	49	0.91	73	
064	0.69	1.30	0.61	1.34	0.01	1.35	85		94	54	1.00	81	
	14	3	2	13	1	1	1		2	2	1	2	
	0.085	0.269	0.048	0.096					26.2	5.3		12.5	
065	0.05	0.04	0.02	0.32	0.06	0.02	1		14	3.6			
066	0.40	0.35	0.20	2.80	0.48	0.17	10		120	31			
	4	4	4	4	2	2	2		4	1			
	0.058	0.055	0.053	0.948	0.118	0.001	1.0		57.0				
067	1.71	0.12	0.16	0.70	0.08	0.07	5.6		328	7		14	0.17
068	1.88	0.13	0.17	0.77	0.08	0.08	6.1		360	7		15	0.19
	20	16	9	14	5	6	6		11	8		6	3
	0.418	0.020	0.018	0.174	0.021	0.042	0.42		335	0.7		2.6	0.097
069	0.22	0.08	0.06	0.44	0.08	0.03	1.1		86	13			
070	1.19	0.42	0.34	2.31	0.46	0.17	6.0		455	67			
	1	1	1	1	1	1	1		1	1			
071	1.14	0.22	0.39	1.95	0.40	0.17	5.3		228	61			
072	1.30	0.25	0.45	2.22	0.46	0.19	6.00		260	69			
	2	2	1	2	1	2	1		1	1			
	0.011	0.006		0.738		0.028							
073	0.25	0.07	0.08	0.50	0.02	0.03						4	
074	1.27	0.35	0.42	2.40	0.12	0.16						20	
	1	1	1	1	1	1						1	
075	1.20	0.30	0.42	2.17	0.12	0.19	8	0.27	419	110		15	0.14
076	1.35	0.33	0.47	2.44	0.13	0.21	9	0.30	470	123		17	0.16
	3	3	3	3	1	3	3	1	4	3		1	1
	0.228	0.060	0.067	0.275		0.010	1.2		211.0	60.9			
077	0.44	0.07	0.10	0.49	0.04	0.03	1.7	0.05	59	10		4	0.03
078	2.26	0.38	0.51	2.49	0.20	0.17	9.0	0.25	300	50		19	0.16
	1	1	1	1	1	1	1	1	1	1		1	1
079	0.26	0.07	0.13	0.51	0.05	0.04	2.6	0.07	79	12		4	0.03
080	1.01	0.27	0.51	1.96	0.20	0.17	10.0	0.25	300	47		16	0.12
	1	1	1	1	1	1	1	1	1	1		1	1
081	0.41	0.06	0.11	0.46		0.05							
082	1.71	0.26	0.48	1.90		0.21							
	8	8	8	8		8							
	0.181	0.029	0.098	0.531		0.014							
083	1.22	0.22	0.34	1.60	0.16	0.15	9.7	0.22	211	95		15	0.14
084	1.38	0.24	0.38	1.81	0.18	0.16	11.0	0.25	238	108		17	0.16
	11	11	7	11	2	2	4	1	8	4		3	1
	0.226	0.055	0.134	0.584	0.038	0.008	12.6		121.0	46.5		17.1	
085	0.41	0.20	0.23	0.75	0.02	0.11	6.5		81	55			
086	0.50	0.25	0.29	0.93	0.03	0.14	8.0		100	68			
	1	1	1	1	1	1	1		1	1			
087	0.46	0.09	0.27	1.39	0.07	0.15	4.0		138	40		17	
088	0.52	0.10	0.31	1.59	0.08	0.17	4.5		158	46		19	
	2	2	2	2	2	1	2		2	1		1	
	0.028		0.106	0.431	0.012		0.7		60.1				
089	0.11	0.04	0.06	0.80	0.07	0.42	6.3		207	5	0.07	4	0.12
090	0.12	0.04	0.07	0.89	0.08	0.47	7.0		230	6	0.08	5	0.13
	2	2	2	2	1	2	1		1	1	1	1	1
	0.001	0.003	0.002	0.021		0.002							

Continues

TABLE 6-1A   Composition of Feeds (Excluding Vitamins) Commonly Used in Horse Diets—*Continued*

Entry No.	Feed Name	International Feed Number[a]	Dry Matter (%)	DE[b] (Mcal/ kg)	(Mcal/ lb)	Crude Protein (%)	Lysine (%)	Ether Extract (%)	Fiber (%)	NDF (%)	ADF (%)	Ash (%)
091	—distillers grains,	5-28-235	92.0	3.21	1.46	27.8	0.81	6.6	11.3	39.6	15.6	3.1
092	dehydrated		100.0	3.49	1.59	30.3	0.88	7.1	12.3	43.0	17.0	3.4
		N	3			4	2	3	4	1	1	2
		SD	2.00			1.20	0.063	3.51	1.67			1.41
093	—ears, ground (corn	4-28-238	86.2	2.83	1.29	7.8	0.17	3.2	8.1	24.1	9.5	1.6
094	and cob meal)		100.0	3.29	1.50	9.0	0.20	3.7	9.4	28.0	11.0	1.9
		N	2			1	1	1	1	1	1	1
		SD	1.13									
095	—grain	4-02-935	88.0	3.38	1.54	9.1	0.25	3.6	2.2	9.5	3.6	1.3
096			100.0	3.84	1.75	10.4	0.28	4.1	2.5	10.8	4.1	1.5
		N	545			527	440	89	72	3	5	64
		SD	2.44			1.42	0.045	0.620	1.17	2.10	0.844	0.306
097	—silage	3-02-912	30.4	0.82	0.37	2.4	0.13	1.0	7.2	14.8	8.4	1.6
098			100.0	2.68	1.22	7.9	0.43	3.3	23.8	48.9	27.8	5.1
		N	8			12	1	5	7	2	8	5
		SD	5.05			0.590		0.044	1.25	3.04	4.24	1.13
	COTTON *Gossypium* spp.											
099	—hulls	1-01-599	90.4	1.71	0.78	3.8		1.5	43.2	79.8	59.0	2.6
100			100.0	1.89	0.86	4.2		1.7	47.8	88.3	65.3	2.9
		N	22			28		26	27	2	4	20
		SD	1.34			0.744		1.19	3.07	2.41	4.31	0.487
101	—seeds, oil residue,	5-01-621	91.0	2.74	1.25	41.3	1.68	1.5	12.2	25.4	17.5	6.5
102	solvent-extracted,		100.0	3.01	1.37	45.4	1.85	1.7	13.4	27.9	19.2	7.1
	ground, 41% protein	N	65			52	18	41	40	3	7	31
		SD	1.36			0.926	0.074	0.566	1.06	1.95	2.85	0.397
	FATS AND OILS											
103	—fat, animal,	4-00-376	99.2	7.94	3.61			98.4				
104	hydrolyzed		100.0	8.00	3.64			99.2				
		N	5					3				
		SD	0.277					1.04				
105	—oil, vegetable	4-05-077	99.8	8.98	4.08			99.7				
106			100.0	9.00	4.09			99.9				
		N	5					6				
		SD	0.289					0.104				
	FESCUE, KENTUCKY 31 *Festuca arundinacea*											
107	—fresh	2-01-902	31.3	0.70	0.32	4.7	0.18	1.7	7.7	19.4	10.8	2.2
108			100.0	2.22	1.01	15.0	0.57	5.5	24.6	62.2	34.4	7.2
		N	5			51	2	18	18	8	8	2
		SD	3.76			2.02	0.077	0.757	2.39	8.36	4.39	3.60
109	—hay, sun-cured, full	1-09-188	91.9	1.89	0.86	11.8		5.1	23.9	61.6	35.7	7.6
110	bloom		100.0	2.06	1.01	12.9		5.6	26.0	67.1	39.2	8.3
		N	3			11		9	8	4	4	9
		SD	3.01			3.35		0.591	2.91	4.51	4.77	1.08
111	—hay, sun-cured,	1-09-189	90.0	1.76	0.80	9.8		4.2	28.1	63.0	35.1	6.1
112	mature		100.0	1.95	0.89	10.8		4.7	31.2	70.0	39.0	6.8
		N	1			13		13	10	1	1	13
		SD				3.58		0.846	2.36			0.920
	FISH, ANCHOVY *Engraulis ringen*											
113	—meal, mechanically	5-01-985	92.0	2.76	1.25	65.5	5.03	4.2	1.0			14.7
114	extracted		100.0	3.00	1.36	71.2	5.47	4.6	1.1			16.0
		N	67			58	40	36	9			47
		SD	1.19			2.24	0.429	1.62	0.015			1.54
	FISH, MENHADEN *Brevoortia tyrannus*											
115	—meal, mechanically	5-02-009	91.7	2.93	1.33	62.2	4.74	9.8	0.7			18.9
116	extracted		100.0	3.20	1.45	67.9	5.17	10.7	0.8			20.6
		N	79			91	64	96	38			87
		SD	1.18			2.65	0.252	1.84	0.206			2.12
	FLAX, COMMON *Linum usitatissimum*											
117	—seeds	5-02-052	93.6	3.40	1.54	21.1	0.86	36.0	6.2		7.5	4.9
118			100.0	3.63	1.65	22.5	0.92	38.4	6.6		8.0	5.2
		N	4			12	1	4	3		1	3
		SD	0.643			2.42		1.48	0.178			1.38
119	—seeds, meal, solvent-	5-02-048	90.2	2.74	1.25	34.6	1.16	1.4	9.1	22.6	17.4	5.9
120	extracted		100.0	3.04	1.38	38.4	1.28	1.6	10.1	25.0	19.0	6.5
		N	25			16	9	18	13	1	1	10
		SD	1.12			1.52	0.082	0.752	0.597			0.257

Entry No.	Macrominerals (%)						Microminerals (mg/kg)						
	Calcium	Phosphorus	Magnesium	Potassium	Sodium	Sulfur	Copper	Iodine	Iron	Manganese	Selenium	Zinc	Cobalt
091	0.10	0.41	0.06	0.17	0.09	0.42	44.2	0.05	205	21	0.24	32	0.08
092	0.10	0.45	0.07	0.18	0.10	0.46	48.0	0.05	223	23	0.27	35	0.09
	2	2	1	1	1	1	1	1	1	1	2	1	1
	0.006	0.433									0.190		
093	0.06	0.23	0.13	0.44	0.02	0.16	6.9	0.03	78	20	0.06	12	0.27
094	0.07	0.27	0.15	0.51	0.02	0.19	8.0	0.03	91	23	0.07	14	0.31
	1	1	2	2	1	2	1	1	1	1	2	1	1
			0.008	0.027		0.035					0.028		
095	0.05	0.27	0.11	0.32	0.03	0.11	3.7		31	5	0.12	19	0.13
096	0.05	0.31	0.12	0.37	0.03	0.13	4.2		35	6	0.14	22	0.14
	48	47	41	35	26	7	33		29	31	17	33	6
	0.070	0.039	0.031	0.033	0.014	0.017	1.3		12.6	1.5	0.125	5.5	0.177
097	0.09	0.07	0.06	0.36	0.02	0.03	1.8		98	11	0.02	13	0.03
098	0.31	0.23	0.20	1.17	0.06	0.12	5.8		321	36.4	0.06	42	0.10
	8	7	9	7	5	2	9		8	8	6	8	1
	0.126	0.034	0.089	0.126	0.052	0.007	1.6		261	17.7	0.035	31.0	
099	0.13	0.08	0.13	0.79	0.02	0.08	12.0		119	108	0.08	20	0.02
100	0.15	0.09	0.14	0.88	0.02	0.08	13.3		131	119	0.09	22	0.02
	16	16	10	11	7	6	4		5	3	1	3	3
	0.016	0.018	0.001	0.055	0.001	0.062	4.0		49.7	2.2		0.1	0.005
101	0.17	1.11	0.54	1.30	0.04	0.26	19.5		188	21	0.90	61	0.48
102	0.18	1.22	0.59	1.40	0.05	0.29	21.4		206	23	0.98	67	0.53
	23	26	14	15	13	6	8		9	10	1	4	4
	0.029	0.100	0.068	0.164	0.007	0.095	2.0		99.5	0.9		4.3	0.785
103													
104													
105													
106													
107	0.16	0.12	0.09	0.72		0.06						7	
108	0.51	0.37	0.27	2.30		0.18						22	
	25	27	24	24		24						1	
	0.105	0.076	0.055	0.489		0.036							
109	0.40	0.29	0.16	2.17		0.24	25.7		174	95		35	
110	0.43	0.32	0.17	2.30		0.26	28.0		190	103		38	
	1	1	1	1		1	1		1	1		1	
111	0.37	0.27	0.14	1.76	0.02		19.8		119	87		32	
112	0.41	0.30	0.16	1.96	0.02		22.0		132	97		35	
	2	2	2	2	1		2		2	2		2	
	0.127	0.070	0.021	0.190			12.7			9.2	22.6	1.4	
113	3.74	2.47	0.25	0.72	0.88	0.72	9.1	3.14	215	11	1.35	105	0.17
114	4.06	2.69	0.27	0.79	0.96	0.78	9.9	3.41	234	12	1.47	114	0.19
	51	52	32	35	32	4	27	2	28	31	27	31	1
	0.537	0.446	0.052	0.277	0.329	0.235	1.8	3.49	63.2	5.9	0.254	16.7	
115	5.01	2.87	0.15	0.71	0.41	0.53	10.3	1.09	545	37	2.15	144	0.15
116	5.46	3.14	0.16	0.77	0.44	0.58	11.3	1.19	594	40	2.34	157	0.17
	68	67	19	21	22	4	20	2	21	21	16	18	2
	0.800	0.314	0.028	0.160	0.131	0.259	3.5	1.41	271	17.7	0.690	19.0	0.068
117	0.22	0.54	0.40	0.74		0.23			90	61			
118	0.23	0.58	0.43	0.79		0.25			100	65			
	2	10	1	10		1			1	1			
	0.003	0.088		0.090									
119	0.39	0.80	0.60	1.38	0.14	0.39	25.7		319	38	0.82		0.19
120	0.43	0.89	0.66	1.53	0.15	0.43	28.5		354	42	0.91		0.21
	12	12	8	8	7	5	6		7	6	1		3
	0.022	0.034	0.007	0.015	0.003	0.026	0.33		12.2	0.15			0.018

Continues

TABLE 6-1A   Composition of Feeds (Excluding Vitamins) Commonly Used in Horse Diets—*Continued*

Entry No.	Feed Name	International Feed Number[a]	Dry Matter (%)	DE[b] (Mcal/kg)	(Mcal/lb)	Crude Protein (%)	Lysine (%)	Ether Extract (%)	Fiber (%)	NDF (%)	ADF (%)	Ash (%)
	LESPEDEZA, COMMON											
	Lespedeza striata											
121	—fresh, late vegetative	2-07-093	25.0	0.55	0.25	4.1		0.5	8.0			3.2
122			100.0	2.20	1.00	16.4		2.0	24.0			12.8
		N	2			1		1	1			1
		SD										
123	—hay, sun-cured,	1-02-554	90.8	1.93	0.88	11.4		2.3	26.2			4.5
124	midbloom		100.0	2.13	0.97	12.6		2.5	30.0			4.9
		N	3			2		2	2			2
		SD	2.93			2.82		0.692	1.30			1.14
	LESPEDEZA, KOBE·											
	Lespedeza striata											
125	—hay, sun-cured,	1-02-574	93.9	1.96	0.89	10.0		2.8	26.2			3.8
126	midbloom		100.0	2.08	0.95	10.6		3.0	27.9			4.1
		N	1			1		1	1			1
		SD										
	MEADOW PLANTS, INTERMOUNTAIN											
127	—hay, sun-cured	1-03-181	95.1	1.60	0.73	8.2		2.4	31.2			8.2
128			100.0	1.69	0.77	8.7		2.5	32.7			8.6
		N	72			219		9	9			4
		SD	1.37			2.25		0.442	1.27			1.04
	MILK											
129	—fresh (horse); see											
130	Appendix Table 1											
131	—skimmed, dehydrated	5-01-175	94.1	3.81	1.73	33.4	2.54	1.0	0.2	0.0	0.0	7.9
132	(cattle)		100.0	4.05	1.84	35.5	2.70	1.0	0.2	0.0	0.0	8.4
		N	57			30	15	27	12	1	1	25
		SD	1.35			2.11	0.249	1.48	0.081			0.398
	MILLET, PEARL											
	Pennisetum glaucum											
133	—fresh	2-03-115	20.7	0.35	0.16	2.1		0.6	6.5			1.9
134			100.0	1.70	0.77	10.1		2.9	31.1			9.2
		N	4			2		2	2			2
		SD	0.190			0.137		0.079	0.092			0.088
135	—hay, sun-cured	1-03-112	87.4	1.34	0.61	7.3		1.8	32.2			8.9
136			100.0	1.53	0.70	8.4		2.0	36.9			10.2
		N	3			2		2	2			2
		SD	0.407			1.00		0.106	1.37			0.155
	MOLASSES AND SYRUP											
137	—beet, sugar, molasses,	4-00-668	77.9	2.65	1.20	6.6		0.2	0.0	0.0	0.0	8.9
138	more than 48%		100.0	3.40	1.55	8.5		0.2	0.0	0.0	0.0	11.4
	invert sugar, more	N	21			12		3				9
	than 79.5 degrees	SD	1.71			1.11		0.105				1.34
	brix											
139	—citrus, syrup (citrus	4-01-241	66.9	2.27	1.03	5.7		0.2	0.0	0.0	0.0	5.1
140	molasses)		100.0	3.40	1.55	8.5		0.3	0.0	0.0	0.0	7.7
		N	12			8		7				5
		SD	3.42			2.18		0.012				1.22
141	—sugarcane, molasses,	4-04-695	94.4	3.21	1.46	9.0		0.8	7.1			12.0
142	dehydrated		100.0	3.40	1.55	9.5		0.9	7.5			12.7
		N	7			7		8	7			4
		SD	2.73			2.36		0.638	3.01			5.74
143	—sugarcane, molasses,	4-04-696	74.3	2.60	1.18	4.3		0.2	0.4		0.3	9.9
144	more than 46%		100.0	3.50	1.59	5.8		0.2	0.5		0.4	13.3
	invert sugar, more	N	84			64		6	1		1	52
	than 79.5 degrees	SD	3.27			2.03		0.240				2.34
	brix (blackstrap)											
	OATS *Avena sativa*											
145	—grain	4-03-309	89.2	2.85	1.30	11.8	0.39	4.6	10.7	24.4	14.2	3.1
146			100.0	3.20	1.45	13.3	0.44	5.2	12.0	27.3	15.9	3.4
		N	97			110	18	108	99	14	2	91
		SD	1.80			1.25	0.061	1.00	1.35	4.52	1.55	0.507
147	—grain, grade 1,	4-18-520	89.0	2.99	1.36	12.5	0.44	5.1	10.8	25.8		3.0
148	heavy, 51.5 kg/hl		100.0	3.36	1.53	14.0	0.49	5.7	12.1	29.0		3.4
		N				35	4	36	31	4		36
		SD				1.56	0.038	0.728	1.29	0.735		0.292
149	—grain, Pacific coast	4-07-999	90.9	2.91	1.32	9.1	0.33	5.0	11.2			3.8
150			100.0	3.20	1.45	10.0	0.37	5.5	12.3			4.2
		N	13			7	4	6	6			3
		SD	0.426			0.377	0.063	0.386	0.508			0.257
151	—groats	4-03-331	89.6	3.09	1.40	15.5	0.55	6.1	2.5			2.0
152			100.0	3.45	1.57	17.3	0.62	6.8	2.8			2.3
		N	41			43	18	35	35			14
		SD	1.55			1.70	0.077	0.924	0.833			0.230

Entry No.	Macrominerals (%)						Microminerals (mg/kg)						
	Calcium	Phosphorus	Magnesium	Potassium	Sodium	Sulfur	Copper	Iodine	Iron	Manganese	Selenium	Zinc	Cobalt
121	0.30	0.07		0.32									
122	1.20	0.28		1.28									
	1	1		1									
123	1.07	0.17	0.22	0.94					282	211			
124	1.18	0.19	0.24	1.04					310	232			
	2	2	2	2					2	2			
	0.076	0.030	0.040	0.024					42.0	113			
125	1.11	0.32	0.27	0.89					291	193			
126	1.18	0.34	0.29	0.95					310	206			
	1	1	1	1					1	1			
127	0.58	0.17	0.16	1.50	0.11								
128	0.60	0.18	0.17	1.58	0.12								
	216	216	209	209	196								
	0.238	0.041	0.088	0.530	0.118								
129													
130													
131	1.28	1.02	0.12	1.60	0.51	0.32	11.7		8	2	0.12	38	0.11
132	1.36	1.09	0.13	1.70	0.54	0.34	12.4		8	2	0.13	41	0.12
	16	14	11	15	8	5	4		2	6	1	4	2
	0.027	0.018	0.014	0.102	0.085	0.004	0.28		2.6	0.1		4.1	0.003
133													
134													
135													
136													
137	0.12	0.02	0.23	4.72	1.16	0.46	16.8		68	4		14	0.36
138	0.15	0.03	0.29	6.06	1.48	0.60	21.6		87	6		18	0.46
	13	11	10	10	8	9	7		8	7		1	5
	0.054	0.007	0.014	0.290	0.079	0.048	1.3		25.2	0.3			0.032
139	1.18	0.09	0.14	0.09	0.28	0.14	72.2		339	26		92	0.11
140	1.76	0.14	0.21	0.14	0.41	0.21	107.9		507	38		137	0.16
	8	7	7	7	5	2	4		4	4		2	2
	0.323	0.035	0.010	0.006	0.029	0.022	3.5		10.9	1.2		0.22	0.002
141	1.03	0.14	0.44	3.39	0.19	0.43	74.9		236	54		31	1.14
142	1.10	0.15	0.47	3.59	0.20	0.46	79.3		250	57		33	1.21
	5	3	1	2	1	1	1		1	1		1	1
	0.347			0.601									
143	0.74	0.08	0.31	2.98	0.16	0.35	48.8	1.56	196	44		15	1.18
144	1.00	0.10	0.42	4.01	0.22	0.47	65.7	2.10	263	59		21	1.59
	32	31	12	16	9	9	8	1	11	11		5	4
	0.182	0.017	0.098	0.885	0.020	0.025	26.0		34.4	6.4		6.0	0.752
145	0.08	0.34	0.14	0.40	0.05	0.21	6.0	0.11	65	36	0.21	35	0.06
146	0.09	0.38	0.16	0.45	0.06	0.23	6.7	0.13	73	40.5	0.24	39	0.06
	64	71	48	47	15	13	27	1	38	37	32	40	8
	0.030	0.048	0.031	0.117	0.026	0.010	1.8		25.2	10.0	0.146	10.9	0.016
147	0.05	0.34	0.12	0.38			5.8		83	37	0.20	35	
148	0.06	0.38	0.14	0.43			6.5		93	41	0.23	39	
	36	36	32	32			26		31	32	32	32	
	0.010	0.066	0.018	0.039			2.7		14.9	7.9	0.220	11.5	
149	0.10	0.31	0.17	0.38	0.06	0.20			73	38	0.08		
150	0.11	0.34	0.19	0.42	0.07	0.22			80	42	0.08		
	3	3	1	2	1	3			1	1	1		
	0.006	0.016		0.007		0.019							
151	0.08	0.42	0.11	0.36	0.03	0.20	6.0	0.11	71	31	0.45	33	
152	0.09	0.47	0.13	0.40	0.04	0.22	6.7	0.12	79	34	0.51	37	
	21	21	12	13	5	6	8	1	9	9	2	2	
	0.029	0.033	0.034	0.038	0.026	0.001	0.9		20.9	7.3	0.233	3.5	

Continues

TABLE 6-1A   Composition of Feeds (Excluding Vitamins) Commonly Used in Horse Diets—*Continued*

Entry No.	Feed Name	International Feed Number[a]	Dry Matter (%)	DE[b] (Mcal/ kg)	(Mcal/ lb)	Crude Protein (%)	Lysine (%)	Ether Extract (%)	Fiber (%)	NDF (%)	ADF (%)	Ash (%)
153	—groats by-product,	1-08-316	91.4	1.66	0.76	4.3	0.15	1.5	29.0			5.8
154	less than 30%		100.0	1.82	0.83	4.7	0.16	1.6	31.7			6.4
	fiber	N	29			27	4	27	27			12
		SD	1.33			1.04	0.062	0.456	2.93			0.451
155	—hay, sun-cured	1-03-280	90.7	1.75	0.79	8.6		2.2	29.1	57.1	34.8	7.2
156			100.0	1.92	0.87	9.5		2.4	32.0	63.0	38.4	7.9
		N	27			32		13	17	1	1	11
		SD	2.55			2.26		0.885	3.57			0.850
157	—hulls	1-03-281	92.4	1.59	0.72	3.8	0.18	1.4	30.6	66.7	36.5	6.1
158			100.0	1.72	0.78	4.1	0.20	1.5	33.2	72.2	39.6	6.6
		N	26			17	4	15	15	4	4	12
		SD	1.14			1.33	0.042	0.813	3.44	5.72	2.06	0.699
159	—straw	1-03-283	92.2	1.49	0.68	4.1		2.0	37.2	68.6	44.2	7.2
160			100.0	1.62	0.74	4.4		2.2	40.4	74.4	47.9	7.8
		N	71			74		16	64	4	5	14
		SD	2.10			1.09		0.424	2.98	2.70	2.48	1.85
	ORCHARDGRASS *Dactylis glomerata*											
161	—fresh, early bloom	2-03-442	23.5	0.54	0.24	3.0		0.9	7.5	12.9	7.2	1.9
162			100.0	2.29	1.04	12.8		3.7	32.0	55.1	30.7	8.1
		N	8			7		5	5	3	2	6
		SD	3.87			2.37		0.801	2.93	8.31	1.98	1.68
163	—fresh, midbloom	2-03-443	27.4	0.55	0.25	2.8		1.0	9.2	15.8	9.8	2.1
164			100.0	2.02	0.92	10.1		3.5	33.5	57.6	35.6	7.5
		N	3			4		2	2	1	1	4
		SD	5.36			3.89		0.368	2.25			0.535
165	—hay, sun-cured,	1-03-425	89.1	1.94	0.88	11.4		2.6	30.2	53.1	30.1	7.6
166	early bloom		100.0	2.17	0.99	12.8		2.9	33.9	59.6	33.8	8.5
		N	7			9		6	5	4	4	6
		SD	3.30			3.51		0.826	1.72	5.28	1.25	1.60
167	—hay, sun-cured, late	1-03-428	90.6	1.72	0.78	7.6		3.1	33.6	58.9	34.2	9.1
168	bloom		100.0	1.90	0.87	8.4		3.4	37.1	65.0	37.8	10.1
		N	7			1		1	1	3	3	3
		SD	1.51							2.77	0.207	3.10
	PANGOLAGRASS *Digitaria decumbens*											
169	—fresh	2-03-493	20.2	0.39	0.10	1.8		0.5	6.6		7.5	1.5
170			100.0	1.95	0.89	9.1		2.3	32.6		36.9	7.6
		N	9			11		9	10		1	9
		SD	2.37			1.34		0.423	1.39			1.12
171	—hay, sun-cured,	1-10-638	91.0	1.72	0.78	9.2		2.2	29.1	63.7	37.1	7.7
172	15–28 days' growth		100.0	1.89	0.86	10.1		2.4	32.0	70.0	40.8	8.5
		N	1			8		2	8	8	8	1
		SD				2.04		0.282	2.57	4.38	3.96	
173	—hay, sun-cured,	1-26-214	91.0	1.62	0.74	6.7		1.8	29.5	66.2	38.1	7.3
174	29–42 days' growth		100.0	1.78	0.81	7.4		2.0	32.4	72.7	41.8	8.0
		N	1			8		2	8	8	8	1
		SD				1.45			3.01	3.73	4.12	
175	—hay, sun-cured,	1-29-573	91.0	1.41	0.64	5.7		1.6	29.3	70.0	41.9	6.9
176	43–56 days' growth		100.0	1.55	0.70	6.3		1.8	32.2	77.0	46.0	7.6
		N	1			8		2	8	8	8	1
		SD				1.79		0.283	3.67	2.33	3.89	
	PEA *Pisum* spp.											
177	—seeds	5-03-600	89.1	3.07	1.40	23.4	1.65	0.9	5.6	17.3	8.0	2.8
178			100.0	3.45	1.57	26.3	1.86	1.0	6.3	19.5	9.0	3.2
		N	19			18	11	14	14	2	1	13
		SD	0.822			3.05	0.136	0.195	0.666	2.64		0.187
	PEANUT *Arachis hypogaea*											
179	—hay, sun-cured	1-03-619	90.7	1.74	0.79	9.9		3.3	30.3		37.2	8.2
180			100.0	1.91	0.87	10.9		3.6	33.4		41.0	9.0
		N	18			14		13	14		1	13
		SD	1.04			0.800		0.963	6.92			1.99
181	—hulls (pods)	1-08-028	90.7	1.13	0.52	7.3	0.36	1.9	57.1	65.1	60.6	4.0
182			100.0	1.25	0.57	8.1	0.40	2.1	63.0	71.7	66.8	4.7
		N	17			18	2	14	15	3	2	8
		SD	1.74			1.51	0.021	1.33	6.72	7.17	1.98	1.84
183	—seeds without coats,	5-03-650	92.4	3.00	1.36	48.9	1.45	2.1	7.7			5.8
184	meal, solvent-		100.0	3.25	1.48	52.9	1.57	2.3	8.4			6.3
	extracted	N	16			12	2	10	10			7
		SD	1.82			3.93	0.062	0.997	1.19			1.02

Entry No.	Macrominerals (%)						Microminerals (mg/kg)						
	Calcium	Phosphorus	Magnesium	Potassium	Sodium	Sulfur	Copper	Iodine	Iron	Manganese	Selenium	Zinc	Cobalt
153	0.15	0.17	0.13	0.50	0.05	0.18	3.0		100	18			
154	0.16	0.19	0.14	0.54	0.06	0.19	3.3		110	20			
	7	8	2	2	2	2	1		1	1			
	0.034	0.050	0.087	0.031	0.017	0.040							
155	0.29	0.23	0.26	1.35	0.17	0.21	4.4		369	90		41	0.07
156	0.32	0.25	0.29	1.49	0.18	0.23	4.8		406	99		45	0.07
	7	26	23	11	16	3	4		5	4		1	3
	0.092	0.059	0.268	0.655	0.060	0.058	1.5		160.7	48.2			0.002
157	0.15	0.14	0.12	0.55	0.06	0.09	6.6		128	25	0.40	27	
158	0.16	0.15	0.13	0.59	0.07	0.10	7.1		138	27	0.43	29	
	9	9	6	8	6	2	4		3	5	1	3	
	0.041	0.048	0.038	0.058	0.088	0.066	3.2		48.4	9.68		8.0	
159	0.22	0.06	0.16	2.33	0.39	0.21	9.5		151	29		5	
160	0.23	0.06	0.17	2.53	0.42	0.22	10.3		164	31		6	
	68	66	18	16	5	6	4		15	5		11	
	0.094	0.038	0.041	0.255	0.072	0.018	0.54		47.1	11.8		1.1	
161	0.06	0.09	0.07	0.80	0.01	0.06	7.8		185	24			
162	0.25	0.39	0.31	3.38	0.04	0.26	33.1		785	104			
	1	1	1	1	1	1	1		2	1			
									21.2				
163	0.06	0.05	0.09	0.57	0.07		13.7		19	37		7	0.03
164	0.23	0.17	0.33	2.09	0.26		50.1		68	136		25	0.10
	1	2	1	1	1		1		1	1		1	1
		0.079											
165	0.24	0.30	0.10	2.59	0.01	0.23	16.9		83	140		36	0.38
166	0.27	0.34	0.11	2.91	0.01	0.26	19.0		93	157		40	0.43
	1	1	1	1	1	1	1		1	1		1	1
167	0.24	0.27	0.10	2.42	0.01		18.1	18.00	76	151	0.03	34	0.27
168	0.26	0.30	0.11	2.67	0.01		20.0	20.00	84	167	0.03	38	0.30
	1	1	1	1	1		1	1	1	1	1	1	1
169	0.08	0.04	0.04	0.29									
170	0.38	0.22	0.18	1.42									
	8	8	5	4									
	0.078	0.064	0.046	0.284									
171	0.53	0.19	0.18	1.55									
172	0.58	0.21	0.20	1.70									
	1	1	1	1									
173	0.42	0.21	0.14	1.27									
174	0.46	0.23	0.15	1.40									
	1	1	1	1									
175	0.35	0.16	0.13	1.00									
176	0.38	0.18	0.14	1.10									
	1	1	1	1									
177	0.12	0.41	0.12	0.95	0.22				64	3		23	
178	0.14	0.46	0.14	1.06	0.25				72	3		26	
	6	7	2	2	2				2	1		1	
	0.054	0.107	0.004	0.117	0.000				22.3				
179	1.12	0.14	0.44	1.25		0.21							0.07
180	1.23	0.16	0.49	1.38		0.23							0.08
	5	5	4	4		1							1
	0.005	0.030	0.065										
181	0.23	0.06	0.14	0.83	0.08	0.09	14.4		278	47		22	0.11
182	0.26	0.07	0.15	0.92	0.09	0.09	15.9		306	51		24	0.12
	6	6	4	5	3	2	2		3	2		2	2
	0.012	0.009	0.045	0.068	0.101	0.008	2.7		15.8	24.5		0.1	0.002
183	0.29	0.61	0.15	1.18	0.03	0.30	15.0	0.06	143	27		33	0.11
184	0.32	0.66	0.17	1.28	0.03	0.33	16.0	0.07	155	29		36	0.12
	2	3	1	2	1	2	1	1	1	1		1	1
	0.247	0.048		0.038		0.002							

Continues

TABLE 6-1A   Composition of Feeds (Excluding Vitamins) Commonly Used in Horse Diets—*Continued*

Entry No.	Feed Name	International Feed Number[a]	Dry Matter (%)	DE[b] (Mcal/kg)	(Mcal/lb)	Crude Protein (%)	Lysine (%)	Ether Extract (%)	Fiber (%)	NDF (%)	ADF (%)	Ash (%)
	PRAIRIE PLANTS, MIDWEST											
185	—hay, sun-cured	1-03-191	91.0	1.48	0.67	5.8		2.1	30.7			7.2
186			100.0	1.62	0.74	6.4		2.3	33.7			8.0
		N	8			5		5	5			4
		SD	1.42			1.63		0.647	1.94			1.07
	REDTOP *Agrostis alba*											
187	—hay, sun-cured, midbloom	1-03-886	92.8	1.83	0.83	11.1		2.4	29.0			6.0
188			100.0	1.97	0.90	12.0		2.6	31.2			6.5
		N	1			1		1	1			1
		SD										
	RICE *Oryza sativa*											
189	—bran with germs	4-03-928	90.5	2.62	1.19	13.0	0.57	13.6	11.7	30.0	18.0	10.4
190			100.0	2.90	1.30	14.4	0.63	15.0	12.9	33.0	20.0	11.5
		N	37			34	19	29	25	8	1	27
		SD	0.744			1.42	0.120	2.14	1.46	6.57		2.16
191	—grain, ground	4-03-938	89.0	3.38	1.54	7.5	0.24	1.6	8.6			5.3
192			100.0	3.80	1.75	8.4	0.28	1.8	9.7			6.0
		N	9			5	4	4	4			2
		SD	0.755			0.604	0.047	0.162	0.597			0.214
193	—hulls	1-08-075	91.9	1.15	0.52	2.8	0.08	1.0	39.2	75.7	66.1	19.0
194			100.0	1.25	0.57	3.1	0.09	1.1	42.7	82.4	72.0	20.6
		N	21			22	3	18	18	3	2	12
		SD	1.45			1.10	0.025	1.07	3.59	4.95	1.54	1.51
195	—mill run	1-03-941	91.6	0.63	0.28	6.3	0.26	5.2	28.9			15.7
196			100.0	0.68	0.31	6.9	0.29	5.6	31.5			17.1
		N	22			23	5	18	20			9
		SD	0.907			0.740	0.061	0.816	2.79			1.53
	RYE, GRAIN *Secale cereale*											
197		4-04-047	87.5	3.36	1.53	12.0	0.41	1.5	2.2	16.3	3.7	1.6
198			100.0	3.84	1.74	13.7	0.47	1.7	2.5	18.6	4.2	1.9
		N	93			70	10	61	64	1	1	66
		SD	1.49			1.05	0.048	0.245	0.595			0.171
	RYEGRASS, ITALIAN *Lolium multiflorum*											
199	—fresh	2-04-073	22.6	0.51	0.23	4.0		0.9	4.7	13.8	8.6	3.9
200			100.0	2.20	1.02	17.9		4.1	20.9	61.0	38.0	17.4
		N	5			2		2	2	1	1	2
		SD	2.35			2.26		0.141	1.27			2.33
201	—hay, sun-cured, late vegetative	1-04-065	85.6	1.57	0.71	8.8		2.1	20.4	54.8	36.0	9.4
202			100.0	1.84	0.83	10.3		2.4	23.8	64.0	42.0	11.0
		N	2			1		1	1	1	1	1
		SD	4.81									
	SORGHUM *Sorghum bicolor*											
203	—grain	4-04-383	90.1	3.21	1.46	11.5	0.26	2.7	2.6	20.7	8.3	1.7
204			100.0	3.56	1.62	12.7	0.29	3.0	2.8	23.0	9.3	1.9
		N	220			230	94	64	43	2	2	59
		SD	2.33			1.97	0.067	0.681	0.989	7.00	0.353	0.445
	SORGHUM, JOHNSON-GRASS *Sorghum halepense*											
205	—hay, sun-cured	1-04-407	90.5	1.50	0.68	6.7		2.0	30.4			7.7
206			100.0	1.66	0.75	7.5		2.2	33.6			8.6
		N	6			2		2	2			2
		SD	0.313			0.349		0.083	0.365			0.343
	SOYBEAN *Glycine max*											
207	—seed coats	1-04-560	90.3	1.69	0.77	11.0	0.47	1.9	36.0	59.9	44.3	4.4
208			100.0	1.88	0.85	12.2	0.53	2.1	39.9	66.3	49.0	4.9
		N	28			27	2	17	23	6	6	10
		SD	3.43			2.51	0.246	0.567	4.79	2.03	2.85	0.481
209	—seeds, meal, solvent-extracted, 44% protein	5-20-637	89.1	3.14	1.43	44.5	2.87	1.4	6.2	13.3	8.9	6.4
210			100.0	3.52	1.60	49.9	3.22	1.6	7.0	14.9	10.0	7.2
		N	119			111	38	87	92	2	3	66
		SD	1.22			1.25	0.242	0.672	0.955	1.27	0.057	0.581
211	—seeds without hulls, meal, solvent-extracted	5-04-612	89.9	3.36	1.53	48.5	3.09	1.0	3.5	6.9	5.5	6.0
212			100.0	3.73	1.70	54.0	3.44	1.1	3.8	7.7	6.1	6.7
		N	78			75	18	41	55	1	3	34
		SD	1.72			1.72	0.168	0.386	0.557		0.750	0.678

Entry No.	Macrominerals (%)						Microminerals (mg/kg)						
	Calcium	Phosphorus	Magnesium	Potassium	Sodium	Sulfur	Copper	Iodine	Iron	Manganese	Selenium	Zinc	Cobalt
185	0.32	0.12	0.24	0.98					80			31	
186	0.35	0.14	0.26	1.08					88			34	
	3	3	2	1					1			1	
	0.008	0.063	0.023										
187	0.58	0.32		1.57									
188	0.63	0.35		1.69									
	1	1		1									
189	0.09	1.57	0.88	1.71	0.03	0.18	11.0		207	358	0.40	30	1.38
190	0.10	1.73	0.97	1.89	0.03	0.20	12.2		229	396	0.44	33	1.53
	21	21	13	18	6	9	6		9	8	1	7	2
	0.064	0.401	0.246	0.220	0.027	0.001	3.8		80.6	125		23.8	0.257
191	0.07	0.32	0.13	0.44	0.06	0.04	2.7	0.04	50	18		15	0.04
192	0.07	0.36	0.14	0.49	0.07	0.05	3.0	0.05	57	20		17	0.05
	4	4	4	5	3	3	1	1	1	1		1	1
	0.021	0.080	0.027	0.359	0.033	0.007							
193	0.11	0.07	0.34	0.60	0.02	0.07	3.1		91	294	0.14	22	
194	0.12	0.07	0.37	0.65	0.02	0.08	3.4		99	320	0.15	24	
	15	14	3	8	1	5	1		1	4	1	1	
	0.067	0.016	0.401	0.622		0.033				27.1			
195	0.15	0.46	0.10	0.52		0.18	9.9		228	9		105	
196	0.17	0.50	0.11	0.57		0.19	10.8		250	10		115	
	10	10	1	2		1	1		1	1		1	
	0.086	0.071		0.333									
197	0.06	0.32	0.11	0.45	0.02	0.15	7.6		63	72	0.38	28	
198	0.07	0.36	0.12	0.51	0.03	0.17	8.6		72	82	0.44	32	
	19	17	10	9	8	8	5		8	5	1	6	
	0.020	0.047	0.039	0.020	0.008	0.008	0.2		25.7	6.8		3.0	
199	0.15	0.09	0.08	0.45	0.00	0.02			226				
200	0.65	0.41	0.35	2.00	0.01	0.10			1,000				
	2	2	1	1	1	1			1				
	0.007	0.007											
201	0.53	0.29		1.34					274				
202	0.62	0.34		1.56					320				
	1	1		1					1				
203	0.04	0.32	0.15	0.37	0.01	0.13	5.4		57	12	0.41	27	0.27
204	0.04	0.36	0.17	0.41	0.01	0.15	6.0		63	14	0.46	30	0.30
	27	26	24	15	14	3	13		23	21	3	21	1
	0.052	0.076	0.049	0.099	0.015	0.035	2.4		19.2	3.8	0.583	22.3	
205	0.80	0.27	0.31	1.22	0.01	0.09			534				
206	0.89	0.30	0.35	1.35	0.01	0.10			590				
	2	2	2	2	1	1			2				
	0.109	0.015	0.004	0.002					14.1				
207	0.48	0.17	0.20	1.17	0.02	0.10	16.1		369	10	0.13	43	0.11
208	0.53	0.18	0.22	1.29	0.03	0.11	17.8		409	11	0.14	48	0.12
	10	8	2	5	4	2	1		2	3	1	2	1
	0.134	0.069	0.070	0.258	0.020	0.028			120	5.0		33.7	
209	0.35	0.63	0.27	1.98	0.03	0.41	19.9		165	31	0.45	50	0.11
210	0.40	0.71	0.31	2.22	0.04	0.46	22.4		185	35	0.51	57	0.12
	26	29	19	21	12	6	15		15	15	10	13	1
	0.109	0.043	0.029	0.245	0.031	0.046	7.9		39.0	3.5	0.281	7.5	
211	0.26	0.64	0.29	2.12	0.01	0.44	20.2	0.11	130	37	0.19	57	0.11
212	0.29	0.71	0.33	2.36	0.01	0.48	22.5	0.12	145	41	0.22	63	0.12
	19	19	6	9	4	2	6	1	2	5	2	7	1
	0.053	0.053	0.026	0.152	0.012	0.003	5.0		35.3	8.66	0.147	7.7	

Continues

TABLE 6-1A   Composition of Feeds (Excluding Vitamins) Commonly Used in Horse Diets—*Continued*

Entry No.	Feed Name	International Feed Number[a]	Dry Matter (%)	DE[b] (Mcal/kg)	(Mcal/lb)	Crude Protein (%)	Lysine (%)	Ether Extract (%)	Fiber (%)	NDF (%)	ADF (%)	Ash (%)
	SUNFLOWER, COMMON *Helianthus annuus*											
213	—seeds without hulls,	5-04-739	92.5	2.59	1.17	45.2	1.68	2.7	11.7	37.0	27.8	7.5
214	meal, solvent-		100.0	2.80	1.27	48.9	1.82	2.9	12.7	40.0	30.0	8.1
	extracted	N	21			22	11	19	20	1	1	14
		SD	1.73			3.96	0.148	0.625	2.18			0.347
	TIMOTHY *Phleum pratense*											
215	—fresh, late vegetative	2-04-903	26.7	0.70	0.29	3.3		1.0	8.6	14.9	7.7	2.0
216			100.0	2.37	1.08	12.2		3.8	32.1	55.7	29.0	7.5
		N	5			8		2	2	6	1	8
		SD	1.86			3.87		0.251	1.93	3.65		0.970
217	—fresh, midbloom	2-04-905	29.2	0.58	0.27	2.7		0.9	9.8			1.9
218			100.0	2.00	0.91	9.1		3.0	33.5			6.6
		N	6			2		2	2			2
		SD	1.76			0.727		0.171	0.262			0.755
219	—hay, sun-cured,	1-04-882	89.1	1.83	0.83	9.6		2.5	30.0	54.7	31.4	5.1
220	early bloom		100.0	2.06	0.94	10.8		2.8	33.6	61.4	35.2	5.7
		N	13			12		10	8	5	5	9
		SD	1.72			3.35		0.545	1.36	1.22	2.38	0.918
221	—hay, sun-cured,	1-04-883	88.9	1.77	0.80	8.6		2.3	30.0	56.6	32.3	5.4
222	midbloom		100.0	1.99	0.90	9.7		2.6	33.8	63.7	36.4	6.1
		N	10			20		11	11	13	13	8
		SD	1.93			1.95		0.344	1.36	2.32	2.19	1.13
223	—hay, sun-cured, full	1-04-884	89.4	1.73	0.78	7.2		2.6	31.5	57.4	33.5	4.6
224	bloom		100.0	1.94	0.88	8.1		2.9	35.2	64.2	37.5	5.2
		N	8			15		7	7	8	8	8
		SD	2.43			1.03		0.729	1.20	2.19	2.27	0.813
225	—hay, sun-cured, late	1-04-885	88.3	1.59	0.72	6.9		2.4	31.5	61.0	36.9	4.8
226	bloom		100.0	1.80	0.82	7.8		2.7	35.6	69.1	41.8	5.4
		N	6			6		6	4	2	2	3
		SD	0.609			0.692		0.360	0.853	1.55	1.76	1.15
	TREFOIL, BIRDSFOOT *Lotus corniculatus*											
227	—fresh	2-20-786	19.3	0.42	0.19	4.0		0.8	4.1	9.0		2.2
228			100.0	2.18	0.99	20.6		4.0	21.2	46.7		11.2
		N	9			12		3	3	11		7
		SD	4.28			3.97		1.30	7.74	11.7		3.25
229	—hay, sun-cured	1-05-044	90.6	1.99	1.19	14.4		1.9	29.3	42.6	32.6	6.7
230			100.0	2.20	1.00	15.9		2.1	32.3	47.5	36.0	7.4
		N	9			8		7	7	1	1	5
		SD	1.46			2.31		0.524	5.32			0.797
	WHEAT *Triticum aestivum*											
231	—bran	4-05-190	89.0	2.94	1.33	15.4	0.56	3.8	10.0	38.1	12.5	5.9
232			100.0	3.30	1.50	17.4	0.63	4.3	11.3	42.8	14.0	6.6
		N	86			64	21	56	54	6	6	37
		SD	1.23			1.13	0.100	0.805	1.28	8.68	1.46	0.606
233	—flour by-product, less	4-05-205	88.9	3.04	1.38	16.4	0.68	4.2	7.8	31.1	10.5	4.6
234	than 9.5% fiber		100.0	3.42	1.55	18.5	0.76	4.7	8.8	35.0	11.8	5.2
		N	65			59	19	48	49	1	2	24
		SD	1.41			1.15	0.084	0.662	1.00		1.13	1.17
235	—fresh, early	2-05-176	22.2	0.64	0.29	6.1		1.0	3.9	10.2	6.3	3.0
236	vegetative		100.0	2.88	1.31	27.4		4.4	17.4	46.2	28.4	13.3
		N	2			2		1	1	1	1	1
		SD	0.988			1.62						
237	—grain, hard, red	4-05-268	88.9	3.43	1.56	13.0	0.40	1.6	2.5	10.4	3.4	1.7
238	winter		100.0	3.86	1.75	14.6	0.45	1.8	2.8	11.7	3.9	2.0
		N	26			22	9	15	13	3	3	14
		SD	1.38			1.35	0.043	0.261	0.377	3.99	1.05	0.249
239	—grain, soft, red	4-05-294	88.4	3.41	1.55	11.4	0.36	1.6	2.4	12.4	3.5	1.8
240	winter		100.0	3.86	1.75	12.9	0.40	1.8	2.7	14.0	4.0	2.1
		N	15			22	2	8	4	1	1	8
		SD	1.68			0.669	0.015	0.208	0.564			0.252
241	—hay, sun-cured	1-05-172	88.7	1.68	0.76	7.7		2.0	25.7	60.3	36.4	7.0
242			100.0	1.90	0.86	8.7		2.2	29.0	68.0	41.0	7.9
		N	12			8		6	9	1	1	4
		SD	3.09			2.22		0.903	2.01			2.05
243	—mill run, less than	4-05-206	89.9	3.12	1.42	15.6	0.57	4.1	8.2		9.9	5.1
244	9.5% fiber		100.0	3.47	1.57	17.3	0.64	4.6	9.1		11.0	5.7
		N	31			17	7	17	18		1	13
		SD	0.740			0.731	0.081	0.319	0.464			0.212

Entry No.	Macrominerals (%)						Microminerals (mg/kg)						
	Calcium	Phosphorus	Magnesium	Potassium	Sodium	Sulfur	Copper	Iodine	Iron	Manganese	Selenium	Zinc	Cobalt
213	0.42	0.94	0.65	1.17	0.03	0.31	3.7		31	19	2.12	97	
214	0.45	1.02	0.70	1.27	0.03	0.33	4.0		33	20	2.30	105	
	11	11	7	7	2	2	1		1	2	1	1	
	0.083	0.248	0.120	0.333	0.006	0.140				6.0			
215	0.11	0.07	0.04	0.73	0.03	0.03	2.4		35	34		10	0.04
216	0.40	0.26	0.16	2.73	0.11	0.13	8.9		132	127		36	0.15
	4	4	4	4	4	2	2		4	2		2	2
	0.123	0.076	0.037	0.401	0.092		2.5		78.2	32.7		7.1	0.082
217	0.11	0.09	0.04	0.60	0.06	0.04	3.3		52	56			
218	0.38	0.30	0.14	2.06	0.19	0.13	11.2		179	192			
	2	2	2	2	2	2	1		2	1			
	0.173	0.067	0.016	0.484	0.000	0.002			29.1				
219	0.45	0.25	0.11	2.14	0.01	0.12	9.8		181	92		55	
220	0.51	0.29	0.13	2.41	0.01	0.13	11		203	103		62	
	3	3	2	2	1	1	1		2	1		1	
	0.077	0.065	0.021	2.10					4.24				
221	0.43	0.20	0.12	1.61	0.01	0.12	14.2		132	50		38	
222	0.48	0.23	0.13	1.82	0.01	0.13	16.0		149	56		43	
	2	2	3	3	1	1	2		3	2		1	
	0.077	0.056	0.046	0.401			15.5		47.5	13.9			
223	0.38	0.18	0.08	1.78	0.07	0.12	25.9		125	83		48	
224	0.43	0.20	0.09	1.99	0.07	0.14	29.0		140	93		54	
	3	4	3	4	3	3	2		2	2		1	
	0.095	0.014	0.038	0.518	0.092	0.010	33.9		24.9	16.9			
225	0.34	0.13	0.08	1.42	0.06	0.14			203				
226	0.38	0.15	0.09	1.61	0.07	0.16			230				
	1	2	1	2	1	1			2				
		0.049		0.035					98.9				
227	0.33	0.05	0.08	0.63	0.02	0.05	2.5		34	16		6	0.09
228	1.74	0.26	0.40	3.26	0.11	0.25	12.8		176	83		31	0.49
	8	8	6	8	6	1	5		5	5		5	6
	0.398	0.046	0.124	1.66	0.053		3.4		125	13.6		7	0.211
229	1.54	0.21	0.46	1.74	0.06	0.23	8.4		206	26		70	0.10
230	1.70	0.23	0.51	1.92	0.07	0.25	9.3		227	29		77	0.11
	3	3	3	4	1	1	1		3	1		1	1
	0.088	0.011	0.200	0.253					149.7				
231	0.13	1.13	0.56	1.22	0.05	0.21	12.6	0.07	145	119	0.51	98	0.07
232	0.14	1.27	0.63	1.37	0.06	0.24	14.2	0.07	163	134	0.57	110	0.08
	30	29	17	17	13	8	8	1	10	8	5	6	3
	0.031	0.211	0.065	0.097	0.019	0.023	1.8		55.7	14.2	0.245	36.0	0.034
233	0.13	0.89	0.34	0.98	0.02	0.17	15.9	0.11	90	114	0.74	97	0.10
234	0.14	1.00	0.38	1.10	0.02	0.19	17.9	0.12	101	128	0.83	109	0.11
	30	31	16	17	5	11	11	1	12	10	1	6	8
	0.047	0.079	0.089	0.047	0.011	0.056	6.0		18.8	3.35		47.3	0.035
235	0.09	0.09	0.05	0.78	0.04	0.05			22				
236	0.42	0.40	0.21	3.50	0.18	0.22			100				
	1	1	1	1	2	2			1				
					0.148	0.035							
237	0.04	0.38	0.13	0.43	0.02	0.13	4.9		36	35	0.25	33	0.14
238	0.05	0.42	0.14	0.48	0.02	0.15	5.5		41	39	0.28	37	0.16
	13	13	11	12	6	3	7		9	9	5	8	4
	0.007	0.045	0.031	0.052	0.023	0.020	1.1		6.7	12.4	0.185	13.1	0.069
239	0.03	0.36	0.12	0.35	0.01	0.13	5.7		40	32	0.04	34	0.10
240	0.04	0.41	0.13	0.40	0.01	0.14	6.5		45	37	0.05	38	0.12
	16	17	16	16	2	15	16		16	16	1	15	1
	0.017	0.027	0.008	0.026	0.001	0.009	1.3		5.6	2.4		2.8	
241	0.13	0.18	0.11	0.88	0.19	0.19			177				
242	0.15	0.20	0.12	0.99	0.21	0.22			200				
	8	8	1	5	2	2			1				
	0.022	0.077		0.440	0.098	0.035							
243	0.10	1.02	0.47	1.20	0.22	0.17	18.5		95	104	0.63		0.21
244	0.11	1.13	0.53	1.33	0.24	0.19	20.6		105	116	0.70		0.23
	16	16	11	10	1	3	5		6	5	1		3
	0.007	0.058	0.093	0.269		0.026	0.5		5.9	3.8			0.027

Continues

TABLE 6-1A   Composition of Feeds (Excluding Vitamins) Commonly Used in Horse Diets—*Continued*

Entry No.	Feed Name	International Feed Number[a]	Dry Matter (%)	DE[b] (Mcal/ kg)	(Mcal/ lb)	Crude Protein (%)	Lysine (%)	Ether Extract (%)	Fiber (%)	NDF (%)	ADF (%)	Ash (%)
245	—straw	1-05-175	91.3	1.48	0.67	3.2		1.8	38.0	72.1	50.2	7.1
246			100.0	1.62	0.74	3.5		2.0	41.7	78.9	55.0	7.7
		N	37			68		15	25	14	16	46
		SD	3.12			1.29		1.10	5.81	4.82	4.95	2.61
	WHEAT, SOFT, WHITE WINTER *Triticum aestivum*											
247	—grain	4-05-337	90.2	3.54	1.61	10.6		1.5	2.2	8.7	2.4	1.5
248			100.0	3.92	1.78	11.8		1.6	2.5	9.7	2.7	1.7
		N	13			16		13	7	12	1	10
		SD	1.05			1.69		0.201	0.359	2.00		0.566
	WHEATGRASS, CRESTED *Agropyron desertorum*											
249	—fresh, early	2-05-420	28.5	0.72	0.33	6.0		0.6	6.2			2.9
250	vegetative		100.0	2.54	1.16	21.0		2.2	21.6			10.0
		N	6			11		22	22			20
		SD	2.68			4.50		0.641	3.10			1.82
	WHEY *Bos taurus*											
251	—dehydrated (cattle)	4-01-182	93.2	3.78	1.72	13.1	0.94	0.7	0.2	0.0	0.0	8.7
252			100.0	4.06	1.84	14.0	1.00	0.8	0.2	0.0	0.0	9.4
		N	68			50	17	38	11	1	1	23
		SD	1.58			2.00	0.147	0.491	0.071			2.03
253	—low lactose,	4-01-186	93.7	3.37	1.53	16.8	1.40	1.0	0.2	0.0	0.0	16.0
254	dehydrated (dried		100.0	3.61	1.64	17.9	1.50	1.1	0.2	0.0	0.0	17.1
	whey product)	N	28			21	15	20	3			14
	(cattle)	SD	1.75			0.958	0.122	0.166	0.119			2.78
	YEAST, BREWERS *Saccharomyces cerevisiae*											
255	—dehydrated	7-05-527	93.1	3.07	1.40	43.4	3.23	1.0	3.2		3.7	6.7
256			100.0	3.30	1.50	46.6	3.47	1.1	3.5		4.0	7.2
		N	65			46	24	32	30		1	26
		SD	1.09			3.11	0.351	0.619	1.11			1.00

NOTE: The following abbreviations were used: DE, digestible energy; NDF, neutral detergent fiber; ADF, acid detergent fiber; CP, crude protein. A blank space indicates that data are not available for that nutrient.

[a] The following variables appear below the feed number: N is the number of observations; SD is the standard deviation.

[b] The following equations were used to calculate DE: for class 1 and 2 feeds, DE (Mcal/kg) $= 4.22 - 0.11\,(\%\ \text{ADF}) + 0.0332\,(\%\ \text{CP}) + 0.00112\,(\%\ \text{ADF})^2$; for class 4 and 5 feeds, DE (Mcal/kg) $= 4.07 - 0.055\,(\%\ \text{ADF})$ (Fonnesbeck, 1981).

Entry No.	Macrominerals (%)						Microminerals (mg/kg)						
	Calcium	Phosphorus	Magnesium	Potassium	Sodium	Sulfur	Copper	Iodine	Iron	Manganese	Selenium	Zinc	Cobalt
245	0.16	0.04	0.11	1.28	0.13	0.17	3.3		143	37		6	0.04
246	0.17	0.05	0.12	1.40	0.14	0.19	3.6		157	41		6	0.05
	51	48	37	39	5	5	34		35	34		30	2
	0.065	0.024	0.024	0.709	0.005	0.005	1.2		39.5	13.7		0.774	0.008
247	0.06	0.30	0.10	0.39	0.02	0.12	7.0		36	36	0.05	27	0.14
248	0.07	0.33	0.11	0.43	0.02	0.13	7.8		40	40	0.05	30	0.15
		10	1	1	1	2	1		1	1	1	1	1
		0.063				0.001							
249	0.12	0.09	0.08										
250	0.44	0.33	0.28										
	22	22	1										
	0.177	0.141											
251	0.85	0.76	0.13	1.16	0.62	1.04	46		194	6	0.06	5	0.11
252	0.92	0.81	0.14	1.25	0.66	1.11	49.9		208	6	0.08	5	0.12
	24	26	11	10	6	8	6		9	8	1	2	4
	0.144	0.128	0.012	0.237	0.272	0.006	4.15		94.8	2.2		2.4	0.002
253	1.50	1.11	0.22	2.86	1.45	1.07	7.0	9.89	245	8	0.05	8	
254	1.60	1.18	0.23	3.05	1.54	1.15	7.5	10.55	262	9	0.06	8	
	15	16	8	10	10	2	1	1	2	3	1	1	
	0.128	0.180	0.079	0.682	0.418	0.056			39.5	0.9			
255	0.14	1.36	0.24	1.68	0.07	0.44	38.4	0.36	83	6	0.91	39	0.51
256	0.15	1.47	0.26	1.81	0.08	0.47	41.3	0.38	89	7	0.98	42	0.54
	26	29	23	21	17	7	15	1	17	17	3	16	4
	0.040	0.149	0.062	0.232	0.029	0.061	29.3		50.0	5.5	0.476	19.1	0.696

TABLE 6-1B Vitamin Composition of Feeds Commonly Used in Horse Diets

Entry No.	Feed Name	International Feed Number[a]	Dry Matter (%)	Carotene (mg/kg)	A Equivalent[b] (IU/kg)	D (IU/kg)	E (mg/kg)	Biotin (mg/kg)	Choline (mg/kg)	Folate (mg/kg)	Niacin (mg/kg)	Pantothenic Acid (mg/kg)	Riboflavin (mg/kg)	Thiamin (mg/kg)	B6 (mg/kg)	B12 (µg/kg)
	ALFALFA *Medicago sativa*															
001	—fresh, late vegetative	2-00-181	23.2			51	39.7									
002			100.0			221	171.5									
		N	14			1	1									
		SD	3.39													
003	—fresh, full bloom	2-00-188	23.8									7.5				
004			100.0									31.3				
		N	8									1				
		SD	3.88													
005	—hay, sun-cured, early bloom	1-00-059	90.5	126.5	50,608	1,806	23.5									
006			100.0	139.8	55,918	2,000	26.0									
		N	43	3		1	1									
		SD	1.92	61.9	24,775											
007	—hay, sun-cured, midbloom	1-00-063	91.0	30.3	41,900	1,810	10.0						9.6			
008			100.0	33.3	46,000	2,000	11.0						10.6			
		N	60	3		1	1						1			
		SD	1.88													
009	—hay, sun-cured, full bloom	1-00-068	90.9	59.1	23,631	1,810	10.0									
010			100.0	65.0	26,000	2,000	11.0									
		N	21	2	2	1	1									
		SD	2.06	56.5	22,627											
011	—meal, dehydrated, 15% protein	1-00-022	90.4	74.5	29,787		81.9	0.25	1,573.1	1.56	41.6	20.7	10.6	2.99	6.28	
012			100.0	82.3	32,934		90.6	0.28	1,739.3	1.73	46.0	22.9	11.7	3.31	6.94	
		N	23	3	3		3	1	1	1	2	2	4	2	1	
		SD	2.18	23.8	9,540		29.8				1.39	0.619	0.256	0.089		
013	—meal, dehydrated, 17% protein	1-00-023	91.8	120.3	48,132		105.9	0.33	1,349.6	4.37	37.0	29.8	12.9	3.39	7.19	
014			100.0	131.1	52,433		115.3	0.36	1,470.2	4.76	40.3	32.4	14.1	3.69	7.83	
		N	72	3	3		7	6	18	12	18	18	14	9	7	
		SD	1.50	37.5	15,013		28.9	0.027	280	2.57	12.5	2.35	1.20	0.196	1.52	
	ALYCECLOVER *Alysicarpus vaginalis*															
015	—hay, sun-cured	1-00-361	89.7													
016			100.0													
		N	4													
		SD	0.901													
	BAHIAGRASS *Paspalum notatum*															
017	—fresh	2-00-464	28.7	52.4	20,973											
018			100.0	182.5	72,994											
		N	9	1	1											
		SD	1.91													
019	—hay, sun-cured	1-00-462	90.0													
020			100.0													
		N	4													
		SD	0.420													
021	—hay, sun-cured, late vegetative	1-20-787	91.0													
022			100.0													
		N	1													
		SD															
023	—hay, sun-cured, early bloom	1-06-138	91.0													
024			100.0													
		N	1													
		SD															
	BARLEY *Hordeum vulgare*															
025	—grain	4-00-549	88.6	2.0	817		23.2	0.15	1,037.1	0.57	78.5	8.1	1.6	4.52	6.48	
026			100.0	2.3	922		26.2	0.17	1,170.9	0.64	88.6	9.1	1.8	5.11	7.32	
		N	237	3	3		10	12	19	13	64	68	68	60	59	
		SD	2.08	1.97	789		14.3	0.039	76.4	0.058	15.8	2.84	0.343	0.697	2.23	

No.	Item	Ref. No.	Stat	1	2	3	4	5	6	7	8	9	10	11	12	13	14
027	—grain, Pacific coast	4-07-939	(as-fed)	88.6	46.4	18,574	975	26.2	0.15	976.3	0.50	46.7	7.1	1.5	4.19	2.89	
028			(dry)	100.0	52.5	21,000	1,102	29.6	0.17	1,101.6	0.56	52.7	8.0	1.7	4.72	3.26	
			N	17	1	1	1	3	1	4	1	5	4	4	2	1	
			SD	0.773				19.2		90.2		6.38	0.410	0.382	0.250		
029	—hay, sun-cured	1-00-495	(as-fed)	88.4													
030			(dry)	100.0													
			N	10													
			SD	2.17													
031	—straw	1-00-498	(as-fed)	91.2													
032			(dry)	100.0													
			N	29													
			SD	3.31													

BEET, SUGAR *Beta vulgaris altissima*

No.	Item	Ref. No.	Stat	1	2	3	4	5	6	7	8	9	10	11	12	13	14
033	—pulp, dehydrated	4-00-669	(as-fed)	91.0	0.2	88	580			820.9		16.8	1.4	0.7	0.39		
034			(dry)	100.0	0.2	97	637			901.9		18.4	1.5	0.8	0.42		
			N	47	2	2	2			6		6	7	7	4		
			SD	1.37	0.000	0.323	4.34			14.4		1.69	0.356	0.232	0.121		

BERMUDAGRASS, COASTAL *Cynodon dactylon*

No.	Item	Ref. No.	Stat	1	2	3	4	5	6	7	8	9	10	11	12	13	14
035	—fresh	2-00-719	(as-fed)	30.3	100.0	39,993											
036			(dry)	100.0	330.5	132,198											
			N	15	4	4											
			SD	6.91	56.6	22,642											
037	—hay, sun-cured, 15–28 days' growth	1-09-207	(as-fed)	88.4													
038			(dry)	100.0													
			N	4													
			SD	0.326													
039	—hay, sun-cured, 29–42 days' growth	1-09-209	(as-fed)	93.0													
040			(dry)	100.0													
041	—hay, sun-cured, 43–56 days' growth	1-09-210	(as-fed)	93.0													
042			(dry)	100.0													
			N	1													

BLUEGRASS, KENTUCKY *Poa pratensis*

No.	Item	Ref. No.	Stat	1	2	3	4	5	6	7	8	9	10	11	12	13	14
043	—fresh, early vegetative	2-00-777	(as-fed)	30.8	148.5	59,420		47.8									
044			(dry)	100.0	481.9	192,770		155.0									
			N	4	1	1		1									
			SD	0.694													
045	—fresh, milk stage	2-00-782	(as-fed)	42.0													
046			(dry)	100.0													
			N	1													
047	—hay, sun-cured, full bloom	1-00-772	(as-fed)	92.1													
048			(dry)	100.0													
			N	2													

BREWERS GRAINS, DEHYDRATED

No.	Item	Ref. No.	Stat	1	2	3	4	5	6	7	8	9	10	11	12	13	14
049		5-02-141	(as-fed)	92.0				26.7	0.63	1,652	0.20	44.0	8.2	1.5	0.6	0.7	0.004
050			(dry)	100.0				29.0	0.68	1,795	0.22	47.8	8.91	1.6	0.7	0.8	0.004

BROME, SMOOTH *Bromus inermis*

No.	Item	Ref. No.	Stat	1	2	3	4	5	6	7	8	9	10	11	12	13	14
051	—fresh, early vegetative	2-00-956	(as-fed)	26.1	152.1	60,854											
052			(dry)	100.0	582.9	233,158											
			N	8	1	1											
			SD	6.39													
053	—fresh, mature	2-08-364	(as-fed)	54.9													
054			(dry)	100.0													
			N	5													
			SD	1.69													
055	—hay, sun-cured, midbloom	1-05-633	(as-fed)	87.6													
056			(dry)	100.0													
			N	2													
057	—hay, sun-cured, mature	1-00-944	(as-fed)	92.6													
058			(dry)	100.0													
			N	6													
			SD	0.539													

Continues

TABLE 6-1B Vitamin Composition of Feeds Commonly Used in Horse Diets—*Continued*

Entry No.	Feed Name	International Feed Number[a]	Dry Matter (%)	Carotene (mg/kg)	A Equivalent[b] (IU/kg)	D (IU/kg)	E (mg/kg)	Biotin (mg/kg)	Choline (mg/kg)	Folate (mg/kg)	Niacin (mg/kg)	Pantothenic Acid (mg/kg)	Riboflavin (mg/kg)	Thiamin (mg/kg)	B6 (mg/kg)	B12 (μg/kg)
059 060	CANARYGRASS, REED *Phalaris arundinacea* —fresh	2-01-113	22.8 100.0 N 4 SD 4.89													
061 062	—hay, sun-cured	1-01-104	89.3 100.0 N 10 SD 2.08	16.9 18.9 N 7 SD 31.5	6,762 7,576 12,631								8.5 9.5 1	3.57 4.00 1		
063 064	CANOLA *Brassica napus-Brassica campestris* —seeds, meal, solvent-extracted	5-06-146	90.8 100.0 N 13 SD 2.41					0.82 0.90 1	6,082.8 6,700.0 1	2.09 2.30 1	183.1 201.6 2 58.8	12.4 13.6 2 5.80	3.0 3.3 2 0.634	4.72 5.20 1		
065 066	CARROT *Daucus* spp. —roots, fresh	4-01-145	11.5 100.0 N 9 SD 1.21	77.9 677.5 N 2 SD 166	31,160 271,000 2 66,468		6.9 60.2 1	0.01 0.07 1		0.14 1.20 1	6.7 58.0 3 12.3	3.5 30.1 1	0.6 4.9 3 1.01	0.67 5.79 3 1.24	1.39 12.05 1	
067 068	CITRUS *Citrus* spp. —pomace without fines, dehydrated	4-01-237	91.1 100.0 N 275 SD 1.52	0.2 0.2 N 2 SD 0.016	85 93 2 6.49				790.0 867.0 7 59.7		22.2 24.4 6 0.342	14.0 15.4 6 1.08	2.1 2.3 7 0.475	1.47 1.61 3 0.134		
069 070	CLOVER, ALSIKE *Trifolium hybridum* —fresh, early vegetative	2-01-314	18.9 100.0 N 3 SD 0.116	72.4 384.0 1	28,977 153,600 1											
071 072	—hay, sun-cured	1-01-313	87.7 100.0 N 9 SD 1.32	163.2 186.0 1	65,285 74,400 1								15.1 17.2 1	4.21 4.80 1		
073 074	CLOVER, LADINO *Trifolium repens* —fresh, early vegetative	2-01-380	19.3 100.0 N 4 SD 1.44	68.1 352.5 1	27,249 141,006 1											
075 076	—hay, sun-cured	1-01-378	89.1 100.0 N 5 SD 2.71	143.7 161.2 1	57,475 64,480 1						9.8 11.0 1	1.0 1.1 1	15.2 17.0 1	3.74 4.20 1		
077 078	CLOVER, RED *Trifolium pratense* —fresh, early bloom	2-01-428	19.6 100.0 N 5 SD 0.464	48.5 247.5 1	19,402 98,994 1											
079 080	—fresh, full bloom	2-01-429	26.2 100.0 N 4 SD 3.00	54.4 207.5 1	21,744 82,994 1											
081 082	—fresh, regrowth, early vegetative	2-28-255	24.0 100.0 N 8 SD 2.58													

No.	Item	Ref. No.	Stat	1	2	3	4	5	6	7	8	9	10	11
083	—hay, sun-cured	1-01-415		88.4	24.3	9,727	1,700		0.09		37.7	9.9	15.7	1.97
084				100.0	27.5	11,005	1,900		0.11		42.6	11.2	17.8	2.22
			N	21	2	2	1		2		2	2	2	2
			SD	1.91	13.1	5,252			0.007		0.571	0.027	0.056	0.033
085	CORN, DENT YELLOW *Zea mays indentata* —aerial part with ears, sun-cured (fodder)	1-28-231		81.0										
086				100.0										
			N											
			SD											
087	—aerial part without ears, without husks, sun-cured (stover, straw)	1-28-233		87.7	3.5									
088				100.0	4.0									
			N	4	1									
			SD	2.43										
089	—cobs, ground	1-28-234		90.1	0.9	361					7.2			
090				100.0	1.0	400					8.0			
			N	3	1	1					1			
			SD	0.246										
091	—distillers' grains, dehydrated	5-28-235		92.0	2.8	1,104			0.48		36.8	11.5	5.2	7.17
092				100.0	3.0	1,200			0.52		40.0	12.5	5.6	7.79
			N	3	1	1			1		1	1	1	2
			SD	2.00										4.36
093	—ears, ground (corn and cob meal)	4-28-238		86.2	3.4	1,379		17.2	0.03		17.2	4.1	0.9	5.95
094				100.0	4.0	1,600		20.0	0.04		20.0	4.8	1.0	6.90
			N	2	1	1		1	1		1	1	1	1
			SD	1.13										
095	—grain	4-02-935		88.0	5.4	2,162		20.9	0.07		22.5	5.1	1.1	6.16
096				100.0	6.1	2,458		23.8	0.08		25.6	5.8	1.2	7.01
			N	545	4	4		2	3		6	5	6	4
			SD	2.44	2.39	958		1.69	0.012		2.43	0.993	0.200	2.27
097	—silage	3-02-912		30.4	13.2	5,285	133				14.2			
098				100.0	43.5	17,406	439				46.6			
			N	8	2	2	1				2			
			SD	5.05	12.0	4,816					5.13			
099	COTTON *Gossypium* spp. —hulls	1-01-599		90.4										
100				100.0										
			N	22										
			SD	1.34										
101	—seeds, oil residue, solvent-extracted, ground, 41% protein	5-01-621		91.0	2.55	2,783.5		14.6	0.56	3.7	40.9	13.7	4.7	5.41
102				100.0	2.81	3,058.0		16.1	0.61	4.1	44.9	15.1	5.2	5.95
			N	65	4	14		4	3	3	14	14	14	4
			SD	1.36	0.333	107		3.13	0.000	0.054	3.23	3.26	0.337	0.771
103	FATS AND OILS —fat, animal, hydrolyzed	4-00-376		99.2										
104				100.0										
			N	5										
			SD	0.277										
105	—oil, vegetable	4-05-077		99.8				57.0						
106				100.0				57.0						
			N	5				1						
			SD	0.289										
107	FESCUE, KENTUCKY 31 *Festuca arundinacea* —fresh	2-01-902		31.3	68.2	27,279								
108				100.0	218.0	87,208								
			N	5	1	1								
			SD	3.76										
109	—hay, sun-cured, full bloom	1-09-188		91.9										
110				100.0										
			N	3										
			SD	3.01										
111	—hay, sun-cured, mature	1-09-189		90.0										
112				100.0										
			N	1										
			SD											

Continues

TABLE 6-1B Vitamin Composition of Feeds Commonly Used in Horse Diets—*Continued*

Entry No.	Feed Name	International Feed Number[a]	Dry Matter (%)	Carotene (mg/kg)	A Equivalent[b] (IU/kg)	D (IU/kg)	E (mg/kg)	Biotin (mg/kg)	Choline (mg/kg)	Folate (mg/kg)	Niacin (mg/kg)	Pantothenic Acid (mg/kg)	Riboflavin (mg/kg)	Thiamin (mg/kg)	B6 (mg/kg)	B12 (µg/kg)
	FISH, ANCHOVY *Engraulis ringen*															
113	—meal, mechanically extracted	5-01-985	92.0				5.0	0.20	4,408.0	0.16	100.0	15.0	7.3	0.52	4.71	352.0
114			100.0				5.4	0.21	4,791.0	0.17	108.7	16.3	8.0	0.57	5.12	382.0
		N	67				3	5	10	5	11	10	10	4	4	8
		SD	1.19				1.96	0.069	720	0.055	29.2	4.19	2.31	0.304	1.48	178
	FISH, MENHADEN *Brevoortia tyrannus*															
115	—meal, mechanically extracted	5-02-009	91.7				5.8	0.18	3,114.5	0.15	54.6	8.6	4.8	0.57	3.81	122.1
116			100.0				6.2	0.20	3,398.0	0.17	59.6	9.4	5.3	0.62	4.15	133.2
		N	79				3	3	5	2	7	7	8	5	3	6
		SD	1.18				2.06	0.076	256	0.070	1.85	0.454	0.230	0.223	2.06	46.4
	FLAX, COMMON *Linum usitatissimum*															
117	—seeds	5-02-052	93.6													
118			100.0													
		N	4													
		SD	0.643													
119	—seeds, meal, solvent-extracted	5-02-048	90.2				7.5		1,393.0	1.26	33.0	14.7	2.9	7.54	8.60	
120			100.0				8.3		1,544.1	1.40	36.5	16.3	3.2	8.36	9.53	
		N	25				3		7	1	9	7	9	6	1	
		SD	1.12				1.57		263		7.56	1.17	0.881	2.98		
	LESPEDEZA, COMMON *Lespedeza striata*															
121	—fresh, late vegetative	2-07-093	25.0													
122			100.0													
		N	2													
		SD														
123	—hay, sun-cured, midbloom	1-02-554	90.8													
124			100.0													
		N	3													
		SD	2.93													
	LESPEDEZA, KOBE *Lespedeza striata*															
125	—hay, sun-cured, midbloom	1-02-574	93.9													
126			100.0													
		N	1													
		SD														
	MEADOW PLANTS, INTERMOUNTAIN															
127	—hay, sun-cured	1-03-181	95.1	31.9	12,744											
128			100.0	33.5	13,393											
		N	72	2	2											
		SD	1.37	13.1	5,251											
	MILK															
129	—fresh (horse)	5-02-401	10.6		42.4					0.00	0.5	2.8	0.2	0.28	0.28	1.3
130			100.0		401					0.01	4.4	26.7	1.8	2.65	2.65	12.1
		N	42		1					1	1	1	1	1	1	1
		SD	0.656													
131	—skimmed, dehydrated (cattle)	5-01-175	94.1			420	9.1	0.33	1,393.2	0.62	11.5	36.4	19.1	3.72	4.09	50.9
132			100.0			446	9.6	0.35	1,479.8	0.66	12.2	38.6	20.3	3.95	4.35	54.1
		N	57			1	3	6	11	6	11	20	22	6	5	5
		SD	1.35				0.743	0.009	47.2	0.023	0.598	4.77	3.37	0.484	0.437	10.0
	MILLET, PEARL *Pennisetum glaucum*															
133	—fresh	2-03-115	20.7	37.8	15,128											
134			100.0	182.5	72,994											
		N	4	1	1											
		SD	0.190													
135	—hay, sun-cured	1-03-112	87.4													
136			100.0													
		N	3													
		SD	0.407													

MOLASSES AND SYRUP

No.	Feed (IFN)	stat	DM	C2	GE	C4	C5	C6	C7	C8	C9	C10	Va	Vb	Vc
137	—beet, sugar, molasses, more than 48% invert sugar, more than 79.5 degrees brix (4-00-668)		77.9	4.0	827.5		41.0	4.5	2.3						
138			100.0	5.1	1,062.4		52.7	5.8	2.9						
		N	21	1	2		6	6	6						
		SD	1.71		116		5.55	0.437	0.249						
139	—citrus, syrup (citrus molasses) (4-01-241)		66.9				26.9	12.7	6.2						
140			100.0				40.3	19.0	9.3						
		N	12				4	4	4						
		SD	3.42				1.58		0.349						
141	—sugarcane, molasses, dehydrated (4-04-695)		94.4	5.2											
142			100.0	5.5											
		N	7	2											
		SD	2.73	0.158											
143	—sugarcane, molasses, more than 46% invert sugar, more than 79.5 degrees brix (blackstrap) (4-04-696)		74.3	5.4	763.4	0.69	36.4	37.4	2.8	0.86	0.11	4.21			
144			100.0	7.3	1,027.1	0.92	49.0	50.3	3.8	1.16	0.15	5.67			
		N	84	2	15	3	19	18	20	7	1	2			
		SD	3.27	0.848	149	0.036	6.81	2.42	0.664	0.081		4.24			

OATS *Avena sativa*

No.	Feed (IFN)	stat	DM	C2	GE	C4	C5	C6	C7	C8	C9	C10	Va	Vb	Vc
145	—grain (4-03-309)		89.2	15.0	967.6	0.27	14.0	9.7	1.4	6.13	0.39	2.61	0.1	44	
146			100.0	16.8	1,084.4	0.30	15.7	10.9	1.6	6.87	0.44	2.93	0.1	49	
		N	97	6	16	4	47	48	49	34	10	29	1	1	
		SD	1.80	8.04	72.6	0.050	3.33	3.69	0.338	0.886	0.085	1.04			
147	—grain, grade 1, heavy, 51.5 kg/hl (4-18-520)		89.0	13.6											
148			100.0	15.3											
		N		32											
		SD		2.60											
149	—grain, Pacific coast (4-07-999)		90.9	20.2	916.9		14.4	11.7	1.2	1.02					
150			100.0	22.2	1,008.7		15.8	12.8	1.3	1.02					
		N	13	1	2		2	2	2	1					
		SD	0.426		12.3		0.100		0.164						
151	—groats (4-03-331)		89.6	14.8	1,131.6	0.20	9.6	13.8	1.2	6.49		1.00			
152			100.0	16.5	1,263.7	0.22	10.7	15.4	1.3	7.25		1.12			
		N	41	1	6	1	9	11	10	6		5			
		SD	1.55		58.1		3.55	2.21	0.163	1.23		0.175			
153	—groats by-product, less than 30% fiber (1-08-316)		91.4		410.6								0.2	87	
154			100.0		449.1								0.2	95	
		N	29		3								1	1	
		SD	1.33		70.7										
155	—hay, sun-cured (1-03-280)		90.7				9.9	3.4	2.8	1.23			27.0	10,792	1,400
156			100.0				10.8	3.7	3.1	1.35			29.7	11,898	1,543
		N	27				6	6	7	3			2	2	1
		SD	2.55				0.877	0.233	1.42	0.538			32.0	12,826	
157	—hulls (1-03-281)		92.4			0.96	9.2	3.1	1.5	0.61		2.19		260.0	
158			100.0			1.04	10.0	3.4	1.7	0.66		2.37		281.3	
		N	26			1	3	2	4	1		1		4	
		SD	1.14				1.58	0.257	0.563					129	
159	—straw (1-03-283)		92.2										3.5	1,401	610
160			100.0										3.8	1,520	661
		N	71										11	815	1
		SD	2.10										2.03		

ORCHARDGRASS *Dactylis glomerata*

No.	Feed (IFN)	stat	DM	C2	GE	C4	C5	C6	C7	C8	C9	C10	Va	Vb	Vc
161	—fresh, early bloom (2-03-442)		23.5												
162			100.0												
		N	8												
		SD	3.87												
163	—fresh, midbloom (2-03-443)		27.4												
164			100.0												
		N	3												
		SD	5.36												
165	—hay, sun-cured, early bloom (1-03-425)		89.1											33.4	13,366	
166			100.0											37.5	15,000	
		N	7											1	1	
		SD	3.30													

TABLE 6-1B Vitamin Composition of Feeds Commonly Used in Horse Diets—*Continued*

Entry No.	Feed Name	International Feed Number[a]	Dry Matter (%)	Carotene (mg/kg)	A Equivalent[b] (IU/kg)	D (IU/kg)	E (mg/kg)	Biotin (mg/kg)	Choline (mg/kg)	Folate (mg/kg)	Niacin (mg/kg)	Pantothenic Acid (mg/kg)	Riboflavin (mg/kg)	Thiamin (mg/kg)	B6 (mg/kg)	B12 (µg/kg)
167	—hay, sun-cured, late bloom	1-03-428	90.6	18.1	7,247											
168			100.0	20.0	8,000											
N			7	1	1											
SD			1.51													
169	PANGOLAGRASS *Digitaria decumbens* —fresh	2-03-493	20.2													
170			100.0													
N			9													
SD			2.37													
171	—hay, sun-cured, 15-28 days' growth	1-10-638	91.0													
172			100.0													
N			1													
173	—hay, sun-cured, 29-42 days' growth	1-26-214	91.0													
174			100.0													
N			1													
175	—hay, sun-cured, 43-56 days' growth	1-29-573	91.0													
176			100.0													
N			1													
177	PEA *Pisum* spp. —seeds	5-03-600	89.1	0.7	285		3.0	0.18	545.8	0.22	30.5	27.7	1.8	4.61	1.96	
178			100.0	0.8	320		3.3	0.20	612.4	0.25	34.3	31.1	2.0	5.17	2.20	
N			19	1	1		1	1	1	1	2	1	2	2	1	
SD			0.822								0.401		1.80	4.53		
179	PEANUT *Arachis hypogaea* —hay, sun-cured	1-03-619	90.7	31.5	12,618	3,260.0							8.8			
180			100.0	34.8	13,911	3,600.0							9.7			
N			18	2	2	1							2			
SD			1.04	21.8	8,759								0.037			
181	—hulls (pods)	1-08-028	90.7	0.8	320											
182			100.0	0.9	353											
N			17	1	1											
SD			1.74													
183	—seeds without coats, meal, solvent-extracted	5-03-650	92.4				2.9	0.33	1,893.8	0.65	177.5	36.8	9.1	5.70	5.95	
184			100.0				3.2	0.36	2,048.9	0.70	192.0	39.8	9.8	6.20	6.43	
N			16				1	1	1	1	1	1	1	1	2	
SD			1.82												2.15	
185	PRAIRIE PLANTS, MIDWEST —hay, sun-cured	1-03-191	91.0			865										
186			100.0			950										
N			8			1										
SD			1.42													
187	REDTOP *Agrostis alba* —hay, sun-cured, midbloom	1-03-886	92.8	4.6	1,856											
188			100.0	5.0	2,000											
N			1	1	1											
189	RICE *Oryza sativa* —bran with germs	4-03-928	90.5				85.3	0.42	1,243.8	1.60	305.9	25.0	2.6	21.94	16.21	
190			100.0				94.2	0.47	1,373.7	1.77	337.9	27.6	2.8	24.23	17.90	
N			37				3	6	8	2	11	10	12	6	2	
SD			0.744				47.6	0.021	78.8	0.925	25.5	7.73	0.142	1.30	4.66	
191	—grain, ground	4-03-938	89.0				14.0	0.80	926.5	0.25	40.0	7.1	0.7	2.90	4.40	
192			100.0				15.7	0.90	1,041.5	0.28	45.0	8.0	0.8	3.20	5.00	
N			9				1	1	3	1	3	2	2	1	1	
SD			0.755						123		9.40	6.02	0.309			

193	—hulls	1-08-075	91.9			7.5	0.62	1,010.0		28.1	7.9	0.5	2.21	0.07
194			100.0			8.1	0.67	1,100.0		30.6	8.6	0.6	2.41	0.08
		N	21			1		1		3	1	2	2	1
		SD	1.45							14.3		0.129	0.023	
195	—mill run	1-03-941	91.6			5.3				36.0		0.6	2.17	
196			100.0			5.8				39.3		0.6	2.37	
		N	22			2				1		1	1	
		SD	0.907			3.08								
	RYE *Secale cereale*													
197	—grain	4-04-047	87.5	0.1	35	14.5	0.05	419.1	0.58	14.1	7.2	1.8	4.51	2.97
198			100.0	0.1	40	16.6	0.06	479.0	0.66	16.1	8.3	2.0	5.16	3.40
		N	93	1	1	4	4	2	5	7	8	9	7	1
		SD	1.49			1.16	0.010	19.1	0.100	1.52	1.42	0.406	1.13	
	RYEGRASS, ITALIAN *Lolium multiflorum*													
199	—fresh	2-04-073	22.6											
200			100.0											
		N	5											
		SD	2.35											
201	—hay, sun-cured, late vegetative	1-04-065	85.6	248.2	99,287									
202			100.0	290.0	116,000									
		N	2	1	1									
		SD	4.81											
	SORGHUM *Sorghum bicolor*													
203	—grain	4-04-383	90.1	1.2	468	10.0	0.26	692.5	0.22	46.6	10.2	1.2	4.52	5.40
204			100.0	1.3	520	12.0	0.29	768.9	0.24	51.8	11.3	1.4	5.02	6.00
		N	220	1	1	1	1	2	1	4	2	5	4	1
		SD	2.33					9.69		14.2	1.77	0.213	0.940	
	SORGHUM, JOHNSONGRASS *Sorghum halepense*													
205	—hay, sun-cured	1-04-407	90.5	35.3	14,102									
206			100.0	38.9	15,579									
		N	6	3	3									
		SD	0.313	8.53	3,414									
	SOYBEAN *Glycine max*													
207	—seed coats	1-04-560	90.3			6.6		586.5		24.8	13.4	3.6	1.59	1.70
208			100.0			7.3		649.4		27.4	14.8	4.0	1.76	1.88
		N	28			1		3		1	1	1	1	1
		SD	3.43					148						
209	—seeds, meal, solvent-extracted,	5-20-637	89.1			3.0	0.36	2,704.6	0.69	26.1	13.8	3.0	6.59	5.90
210	44% protein		100.0			3.4	0.41	3,036.0	0.77	29.3	15.5	3.4	7.40	6.62
		N	119			2	2	12	6	12	12	12	3	3
		SD	1.22			0.007	0.055	75.4	0.161	1.54	0.796	0.412	0.045	1.08
211	—seeds without hulls, meal, solvent-	5-04-612	89.9			3.3	0.32	2,746.5	0.74	21.5	14.8	2.9	3.10	4.92
212	extracted		100.0			3.7	0.36	3,053.9	0.82	23.9	16.4	3.3	3.45	5.47
		N	78			2	2	4	1	6	5	8	5	2
		SD	1.72			0.014	0.003	26.9		0.387	4.09	0.517	2.20	0.255
	SUNFLOWER, COMMON *Helianthus annuus*													
213	—seeds without hulls, meal, solvent-	5-04-739	92.5			11.1		3,627.7		242.1	40.6	3.5	3.10	13.67
214	extracted		100.0			12.0		3,923.3		261.9	43.9	3.8	3.30	14.78
		N	21			2		4		5	3	6	1	2
		SD	1.73			0.298		1,076		66.8	1.14	1.52		3.42
	TIMOTHY *Phleum pratense*													
215	—fresh, late vegetative	2-04-903	26.7	62.6	25,028									
216			100.0	234.0	93,600									
		N	5	1	1									
		SD	1.86											
217	—fresh, midbloom	2-04-905	29.2	56.7	22,672									
218			100.0	194.0	77,600									
		N	6	1	1									
		SD	1.76											
219	—hay, sun-cured, early bloom	1-04-882	89.1	46.8	18,719	11.6								
220			100.0	52.5	21,000	13.0								
		N	13	1	1	1								
		SD	1.72											

Continues

TABLE 6-1B Vitamin Composition of Feeds Commonly Used in Horse Diets—*Continued*

Entry No.	Feed Name	International Feed Number[a]	Dry Matter (%)	Carotene (mg/kg)	A Equivalent[b] (IU/kg)	D (IU/kg)	E (mg/kg)	Biotin (mg/kg)	Choline (mg/kg)	Folate (mg/kg)	Niacin (mg/kg)	Pantothenic Acid (mg/kg)	Riboflavin (mg/kg)	Thiamin (mg/kg)	B6 (mg/kg)	B12 (µg/kg)
221	—hay, sun-cured, midbloom	1-04-883	88.9	47.4	18,964	1,763										
222			100.0	53.3	21,340	1,984										
N			10	1	1	1										
SD			1.93													
223	—hay, sun-cured, full bloom	1-04-884	89.4	42.5	17,000											
224			100.0	47.5	19,000											
N			8	1	1											
SD			2.43													
225	—hay, sun-cured, late bloom	1-04-885	88.3	39.7	15,897											
226			100.0	45.0	18,000											
N			6	1	1											
SD			0.609													
	TREFOIL, BIRDSFOOT *Lotus corniculatus*															
227	—fresh	2-20-786	19.3													
228			100.0													
N			9													
SD			4.28													
229	—hay, sun-cured	1-05-044	90.6	130.6	52,250	1,398							14.6	6.16		
230			100.0	144.2	57,680	1,543							16.1	6.80		
N			9	1	1	1							1	1		
SD			1.46													
	WHEAT *Triticum aestivum*															
231	—bran	4-05-190	89.0	2.6	1,048		14.3	0.38	1,201.1	1.77	196.7	27.9	3.6	8.36	10.33	
232			100.0	2.9	1,177		16.0	0.42	1,349.2	1.98	221.0	31.4	4.0	9.39	11.61	
N			86	3	3		7	4	20	9	15	20	20	7	6	
SD			1.23	0.023	9.37		3.97	0.202	423	0.667	44.8	4.11	1.93	2.13	3.43	
233	—flour by-product, less than 9.5% fiber (middlings)	4-05-205	88.9	3.1	1,227		23.9	0.24	1,227.8	1.24	94.9	17.8	2.0	14.17	9.15	
234			100.0	3.5	1,380		26.9	0.27	1,381.1	1.39	106.8	20.0	2.3	15.94	10.29	
N			65	3	2		5	2	13	7	10	13	11	8	5	
SD			1.41	0.017	7.16		4.89	0.211	360	0.893	4.15	2.14	0.340	1.97	2.73	
235	—fresh, early vegetative	2-05-176	22.2	115.4	46,180						12.6	4.7	6.1			
236			100.0	520.1	208,026						56.9	21.2	27.6			
N			2	1	1						1	1	1			
SD			0.988													
237	—grain, hard, red winter	4-05-268	88.9				11.1	0.11	1,007.2	0.38	53.0	10.2	1.3	4.52	3.02	
238			100.0				12.5	0.12	1,133.2	0.43	59.7	11.4	1.5	5.09	3.40	
N			26				4	1	6	5	7	6	7	5	5	
SD			1.38				2.56		197	0.044	6.09	2.44	0.298	1.33	1.55	
239	—grain, soft, red winter	4-05-294	88.4				15.6		891.7	0.41	53.4	10.1	1.5	4.71	3.21	
240			100.0				17.7		1,008.9	0.46	60.4	11.4	1.7	5.33	3.63	
N			15				1		2	2	2	2	1	2	2	
SD			1.68						186	0.025	5.88	1.97		0.800	2.35	
241	—hay, sun-cured	1-05-172	88.7	75.8	30,304	1,370							15.1			
242			100.0	85.4	34,147	1,543							17.0			
N			12	2	2	1							1			
SD			3.09	37.0	14,803											
243	—mill run, less than 9.5% fiber	4-05-206	89.9				31.9	0.31	1,004.7	1.08	115.6	13.7	2.1	15.25	11.09	
244			100.0				35.5	0.34	1,117.9	1.20	128.7	15.2	2.4	16.97	12.33	
N			31				1	1	9	2	9	9	8	4	2	
SD			0.740						48.5	0.034	10.4	1.08	0.648	0.123	0.122	
245	—straw	1-05-175	91.3	2.0	805	604							2.2			
246			100.0	2.2	882	661							2.4			
N			37	1	1	1							1			
SD			3.12													

	WHEAT, SOFT, WHITE WINTER *Triticum aestivum*														
247	—grain	4-05-337	90.2			18.0	0.11	978.0	0.40	53.0	11.1	1.2	4.70	4.77	
248			100.0			20.0	0.12	1,097.0	0.40	59.0	12.3	1.3	5.30	5.29	
		N	13			1	1	1	1	1	1	1	1	1	
		SD	1.05												
	WHEATGRASS, CRESTED *Agropyron desertorum*														
249	—fresh, early vegetative	2-05-420	28.5	123.4	49,377										
250			100.0	433.6	173,458										
		N	6	1	1										
		SD	2.68												
	WHEY *Bos taurus*														
251	—dehydrated (cattle)	4-01-182	93.2		48.8	0.2	0.35	1,790.7	0.85	10.6	46.2	27.4	4.00	3.22	18.9
252			100.0		52.4	0.2	0.38	1,920.8	0.91	11.4	49.6	29.4	4.29	3.45	20.3
		N	68		1	1	7	11	8	12	26	18	8	6	7
		SD	1.58				0.055	285	0.117	1.13	6.86	3.09	0.474	0.978	4.38
253	—low lactose, dehydrated (dried whey product) (cattle)	4-01-186	93.7				0.51	4,109.2	0.90	17.8	74.7	24.2	5.06	4.49	36.0
254			100.0				0.54	4.387.3	0.96	19.0	79.8	25.8	5.40	4.79	38.4
		N	28				9	11	9	11	11	12	7	6	4
		SD	1.75				0.164	676	0.499	0.979	4.90	12.3	0.337	1.61	3.91
	YEAST, BREWERS *Saccharomyces cerevisiae*														
255	—dehydrated	7-05-527	93.1			2.1	1.04	3,847.0	9.69	443.3	81.5	33.6	85.21	36.66	1.1
256			100.0			2.3	1.12	4,133.9	10.41	476.4	87.6	36.1	91.56	39.40	1.1
		N	65			2	13	21	18	21	31	31	14	16	3
		SD	1.09			0.145	0.123	378	2.49	46.3	25.6	6.05	23.2	8.61	0.141

NOTE: A blank space indicates that data are not available for that nutrient.

[a] The following variables appear below the feed number: *N* = number of observations; SD = standard deviation.

[b] The vitamin A equivalent was calculated as carotene × 400 except values for fresh horse milk (IFN, 5-02-401) and dehydrated whey (IFN, 4-01-182), which represent actual amounts of vitamin A.

TABLE 6-2 Composition of Mineral Supplements for Horses on a 100% Dry Matter Basis

Entry No.	Feed Name	International Feed Number	Dry Matter (%)	Protein Equivalent (%) (N × 6.25)	Macrominerals (%)							Microminerals (ppm or mg/kg)							
					Calcium	Chlorine	Magnesium	Phosphorus	Potassium	Sodium	Sulfur	Cobalt	Copper	Fluorine	Iodine	Iron	Manganese	Selenium	Zinc
01	BONE —charcoal (bone black, bone char)	6-00-402	90	9.4	30.11		0.59	14.14	0.16										
02	—meal, steamed	6-00-400	97	13.2	30.71		0.33	12.86	0.19	5.69	2.51					26,700			100
03	CALCIUM —carbonate, $CaCO_3$	6-01-069	100		39.39		0.05	0.04	0.06	0.06						300	300		
04	—phosphate, monobasic, from defluorinated phosphoric acid	6-01-082	97		16.40		0.61	21.60			1.22	10	10	2,100		15,800	360		90
05	—phosphate, dibasic, from defluorinated phosphoric acid (dicalcium phosphate)	6-01-080	97		22.00		0.59	19.03	0.08	0.06	1.14	10	10	1,800		14,400	300		100
06	—sulfate, dihydrate, $CaSO_4 \cdot 2H_2O$, cp[a]	6-01-089	97		23.28				0.07	0.05	18.62								
07	COLLOIDAL CLAY —clay (soft rock phosphate), see also PHOSPHATE	6-03-947	100[b]		17.00		0.38	9.00			0.10			15,000		19,000	1,000		
08	COBALT —carbonate, $CoCO_3$	6-01-566	99[b]								0.20	460,000				500			
09	COPPER (CUPRIC) —sulfate, pentahydrate, $CuSO_4 \cdot 5H_2O$, cp[a]	6-01-720	100								12.84		254,500						
10	CURACAO phosphate	6-05-586	99[b]		34.34		0.81	14.14			0.20			5,500		3,500			
11	ETHYLENDEDIAMINE dihydroiodide	6-01-842	98[b]												803,400				
12	IRON (FERROUS) —sulfate, heptahydrate	6-20-734	98[b]								12.35					218,400			
13	LIMESTONE —limestone, ground	6-02-632	100		34.00	0.03	2.06	0.02	0.12	0.06	0.04					3,500			
14	—magnesium (dolomitic)	6-02-633	99[b]		22.30	0.12	9.99	0.04	0.36							770			
15	MAGNESIUM —carbonate, $MgCO_3 \cdot Mg(OH)_2$	6-02-754	98[b]		0.02	0.00	30.81	0.02								220			
16	—oxide, MgO	6-02-756	98		3.07		56.20							200			100		

No.	Item	IFN Number	Dry matter (%)	Ca (%)	Mg (%)	P (%)	Cl (%)	K (%)	Na (%)	S (%)	Cu (mg/kg)	I (mg/kg)	Fe (mg/kg)	Mn (mg/kg)	Se (mg/kg)	Zn (mg/kg)
	MANGANESE (MANGANOUS)															
17	—oxide, MnO, cp[a]	6-03-056	99[b]											774,500		
18	—carbonate, $MnCO_3$	6-03-036	97											478,000		
	OYSTERSHELL															
19	—ground (flour)	6-03-481	99	38.00	0.30	0.07		0.10	0.21				2,870	100		
	PHOSPHATE															
20	—defluorinated	6-01-780	100	32.00	0.42	18.00	0.08		4.90		10	20	1,800	6,700	200	60
21	—rock	6-03-945	100	35.00	0.41	13.00	0.06		0.03		10	10	35,000	16,800	200	100
22	—rock, low-fluorine	6-03-946	100	36.00		14.00										
23	—rock, soft: see also CALCIUM	6-03-947	100	17.00	0.38	9.00	0.10						15,000	19,000		1,000
	CALCIUM															
24	—phosphate, monobasic, monohydrate, $NaH_2PO_4 \cdot H_2O$; see also SODIUM	6-04-288	97			22.50			16.68							
	PHOSPHORIC ACID															
25	—H_3PO_4	6-03-707	75	0.05	0.51	31.60	0.02	0.04		1.55	10	10	3,100	17,500	500	130
	POTASSIUM															
26	—bicarbonate, $KHCO_3$, cp[a]	6-29-493	99[b]					39.05								
27	—chloride, KCl	6-03-755	100	0.05	0.34		47.30	50.00	1.00	0.45			600			
28	—iodide, KI	6-03-759	100[b]					21.00				681,700				
29	—sulfate, K_2SO_4	6-06-098	98[b]	0.15	0.61		1.55	41.84	0.09	17.35			710			10
	SELENIUM															
30	—selenite, Na_2SeO_3	6-26-013	98[b]						26.60						456,000	
	SODIUM															
31	—bicarbonate, $NaHCO_3$	6-04-272	100						27.00							
32	—chloride, $NaCl$	6-04-152	100				60.66		39.34							
33	—phosphate, monobasic, monohydrate, $NaH_2PO_4 \cdot H_2O$; see also PHOSPHATE	6-04-288	97			22.50			16.68							
34	—sulfate, decahydrate $Na_2SO_4 \cdot 10H_2O$, cp[a]	6-04-292	97[b]						14.27	9.95						40
35	—tripolyphosphate, $Na_5P_3O_{10}$	6-08-076	96			25.00			31.00							
	ZINC															
36	—oxide, ZnO	6-05-533	100													780,000
37	—sulfate, monohydrate, $ZnSO_4 \cdot H_2O$	6-05-555	99[b]	0.02	0.015					17.68				10		363,600

NOTE: The composition of mineral ingredients that are hydrated (e.g., $CaSO_4 \cdot 2H_2O$) is shown, including the waters of hydration. Mineral compositions of feed-grade mineral supplements vary by source, mining site, and manufacturer. Use manufacturer's analysis when available.

[a]cp = chemically pure.

[b]Dry matter values have been estimated for these minerals.

References

ENERGY

Anderson, C. E., G. D. Potter, J. L. Kreider, and C. C. Courtney. 1983. Digestible energy requirements for exercising horses. J. Anim. Sci. 56:91.

Anderson, M. G. 1975. The effect of exercise on blood metabolite levels in the horse. Eq. Vet. J. 7:27.

Argenzio, R. A., and H. F. Hintz. 1971. Volatile fatty acid tolerance and effect of glucose and VFA on plasma insulin levels in ponies. J. Nutr. 191:723.

Argenzio, R. A., and H. F. Hintz. 1972. Effect of diet on glucose entry and oxidation rates in ponies. J. Nutr. 102:879.

Baker, J. P., and S. G. Jackson. 1983. Concentrate feeds and milk fat of mares. Equine Data Line. September 1. Lexington, Ky.: University of Kentucky.

Bayly, W. M., B. D. Grant, and R. C. Breeze. 1983. The respiratory system in exercise and its possible role in limiting performance. P. 205 in Proc. 8th Eq. Nutr. Physiol. Soc. Symp. Lexington, Ky.: University of Kentucky.

Bowman, V. A., J. P. Fontenot, K. E. Webb, Jr., and T. N. Meacham. 1977. Digestion of fat by equine. P. 40 in Proc. 5th Eq. Nutr. Physiol. Soc. Symp. St. Louis, Mo.

Bowman, V. A., J. P. Fontenot, T. N. Meacham, and K. E. Webb, Jr. 1979. Acceptability and digestibility of animal, vegetable, and blended fats by equine. P. 74 in Proc. 6th Eq. Nutr. Physiol. Symp. College Station, Tex.: Texas A&M University.

Brody, S., ed. 1945. Bioenergetics and Growth. New York: Reinhold Publ. Co.

Burton, J. H., and S. MacNeil. 1985. Trainer feeding practices and nutrient intakes of Standardbreds at three Ontario racetracks. P. S1 in Proc. 9th Eq. Nutr. Physiol. Soc. Symp. East Lansing, Mich.: Michigan State University.

Coffman, J. R., and C. M. Colles. 1983. Insulin tolerance in laminitic ponies. Can. J. Comp. Med. 47:347.

Corke, M. J. 1986. Diabetes mellitus: The tip of the iceberg. Eq. Vet. J. 18:87.

Davison, K. E., G. D. Potter, L. W. Greene, J. W. Evans, and W. C. McMullan. 1987. Lactation and reproductive performance of mares fed added dietary fat during late gestation and early lactation. Pp. 87–92 in Proc. 10th Eq. Nutr. Physiol. Soc. Symp. Fort Collins, Colo.: Colorado State University.

Doreau, M., G. Dussap, and H. Dubroeueq. 1980. Estimation de la production laitière de la jument allaitante par marquage de l'eau corporelle du poulain. Reprod. Nutr. Dev. 20:1883.

Drepper, K., J. O. Gutte, H. Meyer, and F. J. Schway. 1982. Empfehlungen zur Energie und Nahorstoffversong der Pferde. Frankfurt: DLG Verlag.

Duren, S. E., S. G. Jackson, J. P. Baker, and D. K. Aaron. 1987. Effect of dietary fat on blood parameters in excercised Thoroughbred horses. Pp. 674–685 in Equine Exercise Physiology II, J. R. Gillespie and W. E. Robinson, eds. Davis, Calif.: ICEEP Publ.

Ellis, R. N. W., and T. L. J. Lawrence. 1979. Energy and protein under-nutrition in the weanling filly foal. Br. Vet. J. 135:331.

Evans, J. W. 1971. Effect of fasting, gestation, lactation, and exercise on glucose turnover in horses. J. Anim. Sci. 33:1001.

Ford, E. J., and H. A. Simmons. 1985. Gluconeogenesis from cecal propionate in the horse. Br. J. Nutr. 53:55.

Gabel, A. A., D. A. Knight, S. M. Reed, J. A. Pultz, J. D. Powers, L. R. Bramlage, and W. J. Tyznik. 1988. Comparison of incidence and severity of developmental orthopedic disease on 17 farms before and after adjustment of ration. Pp. 163–170 in Proc. 33rd Annu. Conv. Am. Assoc. Eq. Pract. New Orleans, La.

Garcia, M. C., and J. Beech. 1986. Equine intravenous glucose tolerance tests: Glucose and insulin responses of healthy horses fed grain or hay and of horses with pituitary adenomas. Am. J. Vet. Res. 47:570.

Gibbs, P. G., G. D. Potter, R. W. Blake, and W. C. McMullen. 1982. Milk production of Quarter Horse mares during 150 days of lactation. J. Anim. Sci. 54:496.

Glade, M. J. 1983. Nutrition of the racehorse. Eq. Vet. J. 15:31.

Glade, M. J. 1988. The role of endocrine factors in equine developmental orthopedic disease. Pp. 171–190 in Proc. 33rd Annu. Conv. Am. Assoc. Eq. Pract. New Orleans, La.

Glade, M. J., and T. H. Belling, Jr. 1984. Growth plate cartilage metabolism, morphology, and biochemical composition in over- and underfed horses. Growth 48:473.

Glade, M. J., and N. K. Luba. 1987. Serum triiodothyronine and thyroxine concentrations in weanling horses fed carbohydrate by direct gastric infusion. Am. J. Vet. Res. 48:578.

Glade, M. J., and T. J. Reimers. 1985. Effects of dietary energy supply on serum thyroxine, triiodothyronine, and insulin concentrations in young horses. J. Endocr. 104:93.

Goodman, H. M., G. N. VanderNoot, J. R. Trout, and R. L. Squibb. 1973. Determination of energy source utilized by the light horse. J. Anim. Sci. 37:56.

Hambleton, P. L., L. M. Slade, D. W. Hamar, E. W. Kienholz, and L. D. Lewis. 1980. Dietary fat and exercise conditioning effect on metabolic parameters in the horse. J. Anim. Sci. 51:1330.

Henneke, D. R., G. D. Potter, J. L. Kreider, and B. F. Yeates. 1983. Relationship between condition score, physical measurement, and body fat percentage in mares. Eq. Vet. J. 15:371–372.

Henneke, D. R., G. D. Potter, and J. L. Kreider. 1984. Body condition during pregnancy and lactation and reproductive efficiency of mares. Theriogenology 21:897.

Hintz, H. F. 1988. Factors which influence developmental orthopedic disease. Pp. 159–162 in Proc. 33rd Annu. Conv. Am. Assoc. Eq. Pract. New Orleans, La.

Hintz, H. F., R. A. Argenzio, and H. F. Schryver. 1971a. Digestion coefficients, blood glucose levels, and molar percentage of volatile acids in intestinal fluid of ponies fed varying forage–grain ratios. J. Anim. Sci. 33:992.

Hintz, H. F., D. E. Hogue, E. F. Walker, Jr., J. E. Lowe, and H. F. Schryver. 1971b. Apparent digestion in various segments of the digestive tract of ponies fed diets with varying roughage–grain ratios. J. Anim. Sci. 32:245.

Hintz, H. F., M. W. Ross, F. R. Lesser, P. F. Leids, K. K. White, J. E. Lowe, C. E. Short, and H. F. Schryver. 1978. The value of dietary fat for working horses. 1. Biochemical and hematological evaluations. J. Eq. Med. Surg. 2:483.

Hintz, H. F., J. E. Lowe, K. K. White, C. E. Short, and M. Ross. 1982. Nutritional value of fat for exercising horses. Unpublished data, Cornell University.

Householder, P. D., G. D. Potter, and R. E. Lichenwalnes. 1977. Digestible energy and protein requirements of yearling horses. P. 126 in Proc. 5th Eq. Nutr. Physiol. Soc. Symp. St. Louis, Mo.

Jackson, S. G., and J. P. Baker. 1981. Digestible energy requirements of the horse at the canter. P. 141 in Proc. 7th Eq. Nutr. Physiol. Soc. Symp. Warrenton, Va.

Jeffcott, L. B., and J. R. Field. 1985. Current concepts of hyperlipaemia in horses and ponies. Vet. Rec. 116:461.

Jeffcott, L. B., J. R. Field, and J. G. McClean. 1986. Glucose tolerance and insulin sensitivity in ponies and Standardbred horses. Eq. Vet. J. 18:97.

Jordan, R. M. 1979. Energy requirements for lactation of pony mares. P. 27 in Proc. 6th Eq. Nutr. Physiol. Soc. Symp. College Station, Tex.

Kane, E., J. P. Baker, and L. S. Bull. 1979. Utilization of a corn oil supplemented diet by the pony. J. Anim. Sci. 48:1379.

Kline, D. S., and W. W. Albert. 1981. Investigation of a glycogen loading program for Standardbred horses. Pp. 186–194 in Proc. 7th Eq. Nutr. Physiol. Soc. Symp. Warrenton, Va.

Knight, D. A., S. E. Weisbrode, L. M. Schmall, and A. A. Gabel. 1988. Copper supplementation and cartilage lesions in foals. Pp. 191–194 in Proc. 33rd Annu. Conv. Am. Assoc. Eq. Pract. New Orleans, La.

Kronfeld, D. S., and S. Donoghue. 1988. Metabolic convergence in developmental orthopedic disease. Pp. 195–204 in Proc. 33rd Annu. Conv. Am. Assoc. Eq. Pract. New Orleans, La.

Lindholm, A., and K. Piehl. 1974. Fibre composition, enzyme activity, and concentration of metabolites and electrolytes in muscles of Standardbred horses. Acta Vet. Scand. 15:287.

Lindholm, A., H. Bjerneld, and B. Saltin. 1974. Glycogen depletion pattern in muscle fibers of trotting horses. Acta Physiol. Scand. 90:475.

Madigan, J. E., and J. W. Evans. 1973. Insulin turnover and irreversible loss rate in horses. J. Anim. Sci. 36:730.

Mehring, J. S., and W. J. Tyznik. 1970. Equine glucose tolerance. J. Anim. Sci. 30:764.

Meyer, H., and L. Ahlswede. 1976. Über das intrauterine Wachstum und die Korpuzusammenketzung von Fohlen sowie den Nahrstoffkedarf tragender Stuten. Übers. Tiernahig. 4:263.

Meyers, M. C., G. D. Potter, L. W. Greene, S. F. Crouse, and J. W. Evans. 1987. Physiological and metabolic response of exercising horses to added dietary fat. Pp. 107–113 in Proc. 10th Eq. Nutr. Physiol. Soc. Symp. Fort Collins, Colo.: Colorado State University.

Miller, P. A., L. M. Lawrence, and A. M. Hank. 1985. The effect of intense exhaustive exercise on blood metabolites and muscle glycogen levels in horses. Pp. 218–223 in Proc. 9th Eq. Nutr. Physiol. Soc. Symp. East Lansing, Mich.: Michigan State University.

Milligan, J. D., R. J. Coleman, and L. Burwash. 1985. Relationship of energy intake on weight gain in yearling horses. P. 8 in Proc. 9th Eq. Nutr. Physiol. Soc. Symp. East Lansing, Mich.: Michigan State University.

Minieri, L., V. Barbieri, and A. Nizza. 1985. Alcuni aspetti dell' alimentazione del cavallo trottatore in attivita agonistica in compania. Atti Soc. Ital. Sci. Vet. 39:462

Morris, R. P., V. A. Rich, S. L. Ralston, E. L. Squires, and B. W. Pickett. 1987. Follicular activity in transitional mares as affected by body condition and dietary energy. Pp. 93–101 in Proc. 10th Eq. Nutr. Physiol. Soc. Symp. Fort Collins, Colo.: Colorado State University.

Morrison, F. B. 1956. Feeds and Feeding, 21st ed. Ithaca, N.Y.: Morrison Publ. Co.

Mullen, P. A., R. Hopes, and J. Sewell. 1979. The biochemistry, haematology, nutrition, and racing performance of two-year old Thoroughbreds throughout their training and racing season. Vet. Rec. 104:90.

Nash, R., and H. F. Hintz. 1981. Feeding practices at two Standardbred tracks. Hoofbeats 49(9):23.

National Research Council. 1978. Nutrient Requirements of Horses, 4th rev. ed. Washington, D.C.: National Academy Press.

Neuhaus, U. 1959. Milch und Milchgewinnung von Pferdestuten. Z. Tierz. Zuechtungsbiol. 73:370.

Oftedal, O. T., H. F. Hintz, and H. F. Schryver. 1983. Lactation in the horse: Milk composition and intake by foals. J. Nutr. 113:2169.

Ott, E. A., and R. L. Asquith. 1981. Vitamin and mineral supplementation of foaling mares. Pp. 44–53 in Proc. 7th Eq. Nutr. Physiol. Soc. Symp. Warrenton, Va.

Ott, E. A., and R. L. Asquith. 1986. Influence of level of feeding and nutrient content of the concentrate on growth and development of yearling horses. J. Anim. Sci. 62:290.

Pagan, J. D., and H. F. Hintz. 1986a. Composition of milk from pony mares fed various levels of digestible energy. Cornell Vet. 76:139.

Pagan, J. D., and H. F. Hintz. 1986b. Equine energetics. I. Relationship between body weight and energy requirements in horses. J. Anim. Sci. 63:815.

Pagan, J. D., and H. F. Hintz. 1986c. Equine energetics. II. Energy expenditure in horses during submaximal exercise. J. Anim. Sci. 63:822.

Pagan, J. D., H. F. Hintz, and T. R. Rounsaville. 1984. The digestible requirements of lactating pony mares. J. Anim. Sci. 58:1382.

Pagan, J. D., B. Essen-Gustavsson, A. Lindholm, and J. Thornton. 1987a. The effect of exercise and diet on muscle and liver glycogen repletion in Standardbred horses. Pp. 431–435 in Proc. 10th Eq. Nutr. Physiol. Soc. Symp. Fort Collins, Colo.: Colorado State University.

Pagan, J. D., B. Essen-Gustavsson, A. Lindholm, and J. Thornton. 1987b. The effect of dietary energy sources on performance in Standardbred horses. Pp. 686–700 in Equine Exercise Physiology II, J. R. Gillespie and W. E. Robinson, eds. Davis, Calif.: ICEEP Publ.

Platt, H. 1984. Growth of the equine foetus. Eq. Vet. J. 16:247.

Potter, G. D., J. W. Evans, G. W. Webb, and S. P. Webb. 1987. Digestible energy requirements of Belgian and Percheron horses. P. 133 in Proc. 10th Eq. Nutr. Physiol. Soc. Symp. Fort Collins, Colo.: Colorado State University.

Ralston, S. L. 1988. Nutritional management of horses competing in 160 km races. Cornell Vet. 78:53.

Ralston, S. L., and C. A. Baile. 1982. Plasma glucose and insulin concentrations and feeding behavior in ponies. J. Anim. Sci. 54:1132.

Ralston, S. L., G. Van den Brock, and C. A. Baile. 1979. Feed intake patterns and associated blood glucose, free fatty acid, and insulin changes in ponies. J. Anim. Sci. 49:838.

Ralston, S. L., D. E. Freeman, and C. A. Baile. 1983. Volatile fatty acids and the role of the large intestine in the control of feed intake in ponies. J. Anim. Sci. 57:815.

Roberts, M. C. 1975a. Carbohydrate digestion and absorption studies in the equine small intestine. J. S. Afr. Vet. Assoc. 46:19.

Roberts, M. C. 1975b. Carbohydrate digestion and absorption studies in the horse. Res. Vet. Sci. 18:64.

Schougaard, H., J. Falk-Ronne, and J. Philipsson. 1987. Incidence and inheritance of osteochondrosis in the sport horse. In 38th Annual Meeting of the European Association for Animal Production, September 28–October 1, vol. 2, Commissions on cattle production, sheep and goat production, pig production, and horse production, Lisbon, Portugal.

Schryver, H. F., D. W. Meakim, J. E. Lowe, J. Williams, L. V. Soderholm, and H. F. Hintz. 1987. Growth and calcium metabolism in horses fed varying levels of protein. Eq. Vet. J. 19:280.

Scott, B. D., G. D. Potter, J. W. Evans, J. C. Reagor, G. W. Webb, and S. P. Webb. 1987. Growth and feed utilization by yearling horses fed added dietary fat. Pp. 101–106 in Proc. 10th Eq. Nutr. Physiol. Soc. Symp. Fort Collins, Colo.: Colorado State University.

Slade, L. M., L. D. Lewis, C. R. Quinn, and M. L. Chandler. 1975. Nutritional adaptations of horses for endurance performance. Pp.114–128 in Proc. 4th Eq. Nutr. Physiol. Soc. Symp. Pomona, Calif.

Smith, F. H. 1986. Nutritional aspects associated with poor performance in the racehorse. Irish Vet. J. 40:87.

Snow, D. H., and C. J. Vogel. 1987. Equine Fitness: The Care and Training of the Athletic Horse. Devon, England: David and Charles Publ.

Snow, P. H., P. Baxter, and R. J. Rose. 1981. Muscle fibre composition and glycogen depletion in horses competing in an endurance ride. Vet. Rec. 108:374.

Stillions, M. C., and D. Nelson. 1972. Digestible energy during maintenance of the light horse. J. Anim. Sci. 34:981.

Thompson, K. N., S. G. Jackson, and J. R. Rooney. 1987. The effect of above average weight gains on the incidence of radiographic bone aberrations and epiphysitis in growing horses. P. 5 in Proc. 10th Eq. Nutr. Physiol. Soc. Symp. Fort Collins, Colo.: Colorado State University.

Topliff, D. R., G. D. Potter, T. R. Dudson, J. K. Kreider, and G. T. Jessup. 1983. Diet manipulation and muscle glycogen in the equine. Pp. 119–124 in Proc. 8th Eq. Nutr. Physiol. Soc. Symp. Lexington, Ky.: University of Kentucky.

Topliff, D. R., G. D. Potter, J. L. Krieder, T. R. Dutson, and G. T. Jessup. 1985. Diet manipulation, muscle glycogen metabolism and anaerobic work performance in the equine. Pp. 224–229 in Proc. 9th Eq. Nutr. Physiol. Soc. Symp. East Lansing, Mich.: Michigan State University.

Topliff, D. R., S. F. Lee, and D. W. Freeman. 1987. Muscle glycogen, plasma glucose, and free fatty acids in exercising horses fed varying levels of starch. Pp. 421–424 in Proc. 10th Eq. Nutr. Physiol. Soc. Symp. Fort Collins, Colo.: Colorado State University.

Vermorel, M., R. Jarrige, and W. Martin-Rosset. 1984. In Le Cheval, J. Jarrige and W. Martin-Rosset, eds. Paris: Institut National de la Recherche Agronomique.

Webb, S. P., G. D. Potter, and J. W. Evans. 1987. Physiologic and metabolic response of race and cutting horses to added dietary fat. P. 115 in Proc. 10th Eq. Nutr. Physiol. Soc. Symp. Fort Collins, Colo.: Colorado State University.

Willard, J. C., L. S. Bull, and J. P. Baker. 1978. Digestible energy requirements of the light horse at two levels of work. P. 324 in Proc. 70th Annu. Meet. Am. Soc. Anim. Sci. East Lansing, Mich.: Michigan State University.

Winter, L. D., and H. F. Hintz. 1981. Feeding practices at two Throughbred racetracks. P. 136 in Proc. 7th Eq. Nutr. Physiol. Soc. Symp. Warrenton, Va.

Wolter, R., J. P. Valette, and J. M. Marion. 1986. Magnesium et effort d'endurance chez le poney. Ann. Zootech. 35:255

Zimmerman, R. A. 1981. Energy needs of lactating mares. P. 127 in Proc. 7th Eq. Nutr. Physiol. Soc. Symp. Warrenton, Va.

PROTEIN

Banach, M., and J. W. Evans. 1981. Effects of inadequate energy during gestation and lactation on the estrous cycle and conception rates of mares and on their foal weights. P. 97 in Proc. 7th Eq. Nutr. Physiol. Soc. Symp. Warrenton: Va.

Bergin, W. C., H. T. Gier, R. A. Frey, and G. B. Marion. 1967. Development horizons and measurements useful for age determination of equine embryos and fetuses. P. 179 in Proc. Am. Assoc. Eq. Pract. New Orleans, La.

Boren, S. R., D. R. Topliff, D. W. Freeman, R. J. Bahr, D. G. Wagner, and C. V. Maxwell, 1987. Growth of weanling Quarter Horses fed varying energy and protein levels. P. 43 in Proc. 10th Eq. Nutr. Physiol. Soc. Symp. College Station, Tex.: Texas A&M University.

Borton, A., D. L. Anderson, and S. Lyford. 1973. Studies of protein quality and quantity in the early weaned foal. P. 19 in Proc. 3rd Eq. Nutr. Physiol. Soc. Symp. Gainesville, Fla.: University of Florida.

Breuer, L. H., and D. L. Golden. 1971. Lysine requirement of the immature equine. J. Anim. Sci. 33:227.

Breuer, L. H., L. H. Kasten, and J. D. Word. 1970. Protein and amino acid utilization in the young horse. P. 16 in Proc. 2nd. Eq. Nutr. Physiol. Soc. Symp. Ithaca, N.Y.: Cornell University.

Ellis, R. N. W., and T. L. J. Lawrence. 1979. Energy and protein under nutrition in the weanling filly foal. Br. Vet. J. 135:331.

Freeman, D. W., G. D. Potter, G. T. Schelling, and J. L. Kreider. 1985. Nitrogen metabolism in the mature physically conditioned horse. II. Response to varying nitrogen intake. P. 236 in Proc. 9th. Eq. Nutr. Physiol. Soc. Symp. East Lansing, Mich.: Michigan State University.

Freeman, D. W., G. D. Potter, G. T. Schelling, and J. L. Kreider. 1988. Nitrogen metabolism in mature horses at varying levels of work. J. Anim. Sci. 66:407.

Gibbs, P. G., G. D. Potter, R. W. Blake, and W. C. McMullan. 1982. Milk production of Quarter Horse mares during 150 days of lactation. J. Anim. Sci. 54:496.

Gibbs, P. G., D. H. Sigler, and T. B. Goehring, 1987. Influence of diet on growth and development of yearling horses. P. 37 in Proc. 10th Eq. Nutr. Physiol. Soc. Symp. Fort Collins, Colo.: Colorado State University.

Gibbs, P. G., G. D. Potter, G. T. Schelling, J. L. Kreider, and C. L. Boyd. 1988. Digestion of hay protein in different segments of the equine digestive tract. J. Anim. Sci. 66:400.

Gill, R. J., G. D. Potter, J. L. Kreider, G. T. Schelling, W. L. Jenkins, and K. K. Hines. 1985. Nitrogen status and postpartum pH levels of mares fed varying levels of protein. P. 84 in Proc. 9th. Eq. Nutr. Physiol. Soc. Symp. East Lansing, Mich.: Michigan State University.

Glade, M. J. 1983. Nutrition and performance of racing Thoroughbreds. Eq. Vet. J. 15:31.

Glade, M. J., D. Beller, J. Bergen, D. Berry, E. Blonder, J. Bradley, M. Cupelo, and J. Dallas. 1985. Dietary protein in excess of requirements inhibits renal calcium and phosphorus reabsorption in young horses.Nutr. Rep. Int. 31:649.

Godbee, R. G., and L. M. Slade. 1981. The effect of urea or soybean meal on the growth and protein status of young horses. J. Anim. Sci. 53:670.

Harvey, A. L., B. H. Thomas, C. C. Culbertson, and E. V. Collins. 1939. The effect of limited feeding of oats and timothy hay during work on the nitrogen balance of draft geldings. P. 94 in Proc. Am. Soc. Anim. Prod. Chicago, Ill.

Hinkle, D. K., G. D. Potter, J. L. Kreider, G. T. Schelling, and J. G. Anderson. 1981. Nitrogen balance in exercising mature horses fed varying levels of protein. P. 91 in Proc. 7th Eq. Nutr. Physiol. Soc. Symp. Warrenton, Va.

Hintz, H. F., and H. F. Schryver. 1972. Nitrogen utilization in ponies. J. Anim. Sci. 34:592.

Hintz, H. F., J. E. Lowe, and H. F. Schryver. 1969. Protein sources for horses. P. 65 in Proc. Cornell Nutr. Conf. Ithaca, N.Y.: Cornell University Press.

Hintz, H. F., J. E. Lowe, A. J. Clifford, and W. J. Visek. 1970. Ammonia intoxication resulting from urea ingestion by ponies. J. Am. Vet. Med. Assoc. 157:963.

Hintz, H. F., H. F. Schryver, and J. E. Lowe. 1971. Comparison of a blend of milk products and linseed meal as protein supplements for young growing horses. J. Anim. Sci. 33:1274.

Hintz, H. F., K. K. White, C. E. Short, J. E. Lowe, and M. Ross. 1980. Effects of protein levels on endurance horses. P. 202 in Proc. 72nd Annu. Meet. Am. Soc. Anim. Sci. (Abstract).

Holtan, D. W., and L. D. Hunt. 1983. Effect of dietary protein on reproduction in mares. P. 107 in Proc. 8th Eq. Nutr. Physiol. Soc. Symp. Lexington, Ky.: University of Kentucky.

Jordan, R. M. 1979. Energy requirements for lactation of pony mares. P. 27 in Proc. 6th Eq. Nutr. Physiol. Soc. Symp. College Station, Tex.: Texas A&M University.

Meakim, D. W. 1979. Bone mineral content determination of the equine third metacarpal via radiographic photometry and the effect of dietary lysine and methionine supplementation on growth and bone development in the weanling foal. M.S. thesis, University of Florida.

Meakim, D. W., H. F. Hintz, H. F. Schryver, and J. E. Lowe. 1981. The effect of dietary protein on calcium metabolism and growth of the weanling foal. P. 95 in Proc. Cornell Nutr. Conf. Ithaca, N.Y.: Cornell University Press.

Meyer, H. 1987. Nutrition of the equine athlete. Pp. 650–657 in Equine Exercise Physiology II, J. R. Gillespie and W. E. Robinson, eds. Davis, Calif.: ICEEP Publ.

Meyer, H., and L. Ahlswede. 1978. The intrauterine growth and body composition of foals and the nutrient requirements of pregnant mares. Anim. Res. Dev. 8:86.

Meyer, H., S. vom Stein, and M. Schmidt. 1985. Investigations to determine endogenous faecal and renal N losses in horses. P. 68 in Proc. 9th Eq. Nutr. Physiol. Soc. Symp. East Lansing, Mich.: University of Michigan.

Meyers, M. C., G. D. Potter, L. W. Greene, S. F. Crouse, and J. W. Evans. 1987. Physiological and metabolic response of exercising horses to added fat. P. 107 in Proc. 10th Eq. Nutr. Physiol. Soc. Symp. Fort Collins, Colo.: Colorado State University.

Nelson, D. D., and W. J. Tyznik. 1971. Protein and nonprotein nitrogen utilization in the horse. J. Anim. Sci. 32:68.

Oftedal, O. T., H. F. Hintz, and H. F. Schryver. 1983. Lactation in the horse: Milk composition and intake by foals. J. Nutr. 113:2196.

Ott, E. A., and R. L. Asquith. 1983. Influence of protein and mineral intake on growth and bone development of weanling horses. P. 39 in Proc. 8th Eq. Nutr. Physiol. Soc. Symp. Lexington, Ky.: University of Kentucky.

Ott, E. A., and R. L. Asquith. 1986. Influence of level of feeding and nutrient content of the concentrate on growth and development of yearling horses. J. Anim. Sci. 62:290.

Ott, E. A., R. L. Asquith, J. P. Feaster, and F. G. Martin. 1979. Influence of protein level and quality on the growth and development of yearling foals.J. Anim. Sci. 49:620.

Ott, E. A., R. L. Asquith, and J. P. Feaster. 1981. Lysine supplementation of diets for yearling horses. J. Anim. Sci. 53:1496.

Patterson, P. H., C. N. Coon, and I. M. Hughes. 1985. Protein requirements of mature working horses. J. Anim. Sci. 61:187.

Platt, H. 1984. Growth of the equine foetus. Eq. Vet. J. 16:247.

Potter, G. D. 1981. Use of cottonseed meal in rations for young horses. Feedstuffs (December 28):29.

Potter, G. D., and J. D. Huchton. 1975. Growth of yearling horses fed different sources of protein with supplemental lysine. P. 19 in Proc. 4th Eq. Nutr. Physiol. Soc. Symp. Pomona, Calif.

Prior, R. L., H. F. Hintz, J. E. Lowe, and W. J. Visek. 1974. Urea recycling and metabolism of ponies. J. Anim. Sci. 8:565.

Quinn, C. R. 1975. Isobutylidene diurea as a protein substitute for horses. Ph.D. dissertation, Colorado State University.

Ralston, S. L. 1988. Nutritional management of horses competing in 160 km races. Cornell Vet. 78:53.

Ratliff, F. D., R. K. King, and J. B. Reynolds. 1963. The effect of urea in rations for horses. Vet. Med. 58:945.

Reitnour, C. M. 1978. Response to dietary nitrogen in ponies. Vet. J. 10:65.

Reitnour, C. M., and R. L. Salsbury. 1976. Utilization of proteins by the equine species. Am. J. Vet. Res. 37:1065.

Reitnour, C. M., and J. M. Treece. 1971. Relationship of nitrogen source to certain blood components and nitrogen balance in the equine. J. Anim. Sci. 32:487.

Robb, J., R. B. Harper, H. F. Hintz, J. E. Lowe, J. T. Reid, and H. F. Schryver. 1972. Body composition of the horse: Interrelationships among proximate components, energy content, liver and kidney size, and body size and energy value of protein and fat. Anim. Prod. 14:25.

Rusoff, L. L., R. B. Lank, T. E. Spillman, and H. B. Elliot. 1965. Nontoxicity of urea feeding to horses. Vet. Med. 60:1123.

Scott, B. D., G. D. Potter, J. W. Evans, J. C. Reagor, G. W. Webb, and S. P. Webb. 1987. Growth and feed utilization by yearling horses fed added dietary fat. P. 101 in Proc. 10th Eq. Nutr. Physiol. Soc. Symp. Fort Collins, Colo.: Colorado State University.

Slade, L. M., D. W. Robinson, and K. E. Casey. 1970. Nitrogen metabolism in nonruminant herbivores. II. Comparative aspects of protein digestion. J. Anim. Sci. 30:761.

Slade, L. M., L. D. Lewis, C. R. Quinn, and M. L. Chandler. 1975. Nutritional adaptation of horses for endurance type performance. P. 114 in Proc. 4th Eq. Nutr. Physiol. Soc. Symp. Pomona, Calif.

Szcurek, E. M., S. G. Jackson, J. R. Rooney, and J. P. Baker. 1987. Influence of confinement, plane of nutrition and low heel on the occurrence of acquired, forelimb contracture. P. 19 in Proc. 10th Eq. Nutr. Physiol. Soc. Symp. Fort Collins, Colo.: Colorado State University.

Teeter, S. M., M. C. Stillions, and W. E. Nelson, 1967. Recent observations on the nutrition of mature horses. P. 39 in Proc. Am. Assoc. Eq. Pract., December 1967. New Orleans, La.

Ullrey, D. E., R. D. Struthers, D. G. Hendricks, and B. E. Brent. 1966. Composition of mare's milk. J. Anim. Sci. 25:217.

Webb, S. P., G. D. Potter, and J. W. Evans. 1987. Physiologic and metabolic response of race and cutting horses to added dietary fat. P. 115 in Proc. 10th Eq. Nutr. Physiol. Soc. Symp. Fort Collins, Colo.: Colorado State University.

Wirth, B. L. 1977. Cottonseed meal and lysine for weanling foals. M.S. thesis. Texas A&M University, College Station, Tex.

Yoakam, S. C., W. W. Kirkham, and W. M. Beeson. 1978. Effect of protein level on growth in young ponies. J. Anim. Sci. 46:983.

Zimmerman, R. A. 1985. Effect of ration on composition of mare's milk. P. 96 in Proc. 9th Eq. Nutr. Physiol. Soc. Symp. East Lansing, Mich.: Michigan State University.

MINERALS

Macrominerals

Baucus, K. L., S. L. Ralston, V. A. Rich, and E. L. Squires. 1987. The effect of dietary copper and zinc supplementation on composition of mare's milk. Pp. 179–184 in Proc. 10th Eq. Nutr. Physiol. Soc. Symp. Fort Collins, Colo.: Colorado State University.

Blaney, B. J., R. J. W. Gartner, and R. A. McKenzie. 1981. The inability of horses to absorb calcium oxalate. J. Agr. Sci. Camb. 97:639.

Caple, I. W., J. M. Bourke, and P. G. Ellis. 1982. An examination of the calcium and phosphorus nutrition of Thoroughbred racehorses. Aust. Vet. J. 58:132.

Corke, M. J. 1981. An outbreak of sulphur poisoning in horses. Vet. Rec. 109:212.

Drepper, K., J. O. Gutte, H. Meyer, and F. J. Schwarz. 1982. Energie- und Nahrstoffbedarf landwirtschaftlicher Nutztiere. Nr. 2 Empfehlungen zur Energie- und Nahrstoffversorgung der Pferde. Frankfurt am Main, Germany: DLG Verlag.

El Shorafa, W. M., J. P. Feaster, and E. A. Ott. 1979. Horse metacarpal bone: Age, ash content, cortical area, and failure-stress interrelationships. J. Anim. Sci. 49:979.

Fettman, M. J., L. E. Chase, J. Bentinck-Smith, C. E. Coppock, and S. A. Zinn. 1984. Effects of dietary chloride restriction in lactating dairy cows. J. Am. Vet. Med. Assoc. 185:167.

Green, H. H., et al. 1935. Hypomagnesaemia in equine transit tetany. J. Comp. Pathol. 48:74.

Harrington, D. D. 1974. Pathologic features of magnesium deficiency in young horses fed purified rations. Am. J. Vet. Res. 35:503.

Harrington, D. D., and J. J. Walsh. 1980. Equine magnesium supplements: Evaluation of magnesium sulphate and magnesium carbonate in foals fed purified diets. Eq. Vet. J., 12:32.

Hintz, H. F., and H. F. Schryver. 1972. Magnesium metabolism in the horse. J. Anim. Sci. 35:755.

Hintz, H. F., and H. F. Schryver. 1973. Magnesium, calcium, and phosphorus metabolism in ponies fed varying levels of magnesium. J. Anim. Sci. 37:927.

Hintz, H. F., and H. F. Schryver. 1976. Potassium metabolism in ponies. J. Anim. Sci. 42:637.

Hintz, H. F., A. J. Williams, J. Rogoff, and H. F. Schryver. 1973. Availability of phosphorus in wheatbran when fed to ponies. J. Anim. Sci. 36:522.

Hintz, H. F., H. F. Schryver, J. Doty, C. Lakin, and R. A. Zimmerman. 1984. Oxalic acid content of alfalfa hays and its influence on the availability of calcium, phosphorus, and magnesium to ponies. J. Anim. Sci. 58:939.

Jarrige, R., and W. Martin-Rosset. 1981. Le cheval: Reproduction, selection, alimentation, exploitation. XIII Journées du Grenier de Theix. Paris: Institut National de la Recherche Agronomique.

Jordan, R. M., V. S. Meyers, B. Yoho, and F. A. Spurrell. 1975. Effect of calcium and phosphorus levels on growth, reproduction, and bone development of ponies. J. Anim. Sci. 40:78.

Kato, G., J. S. Kelly, K. Drnjevic, and G. Somjen. 1968. Anaesthetic action of magnesium ions. Can. Anaesth. Soc. J. 15:539.

Knight, D. A., A. A. Gabel, S. M. Reed, L. R. Bramlage, W. J.

Tyznik, and R. M. Embertson. 1985. Correlation of dietary mineral to incidence and severity of metabolic bone disease in Ohio and Kentucky. P. 445 in Proc. 31st Am. Assoc. Eq. Pract., F. J. Milne, ed. Lexington, Ky.: Am. Assoc. Eq. Pract.

Krook, L., and J. E. Lowe. 1964. Nutritional secondary hyperparathyroidism in the horse. Pathol. Vet. 1(Suppl. 1):1.

Krook, L., and G. A. Maylin. 1988. Fractures in Thoroughbred race horses. Cornell Vet. 98(Suppl. 11):1.

Lloyd, K., H. F. Hintz, J. D. Wheat, and H. F. Schryver. 1987. Enteroliths in horses. Cornell Vet. 77:172.

McKenzie, R.A., B. J. Blaney, and R. J. W. Gartner. 1981. The effect of dietary oxalate on calcium, phosphorus and magnesium balances in horses. J. Agr. Sci. Camb. 97:69.

Meyer, H. 1979. Magnesium stoffwechsel und magnesium bedarf des pferdes obers. Z. Tierernährg Band. 7:75.

Meyer, H. 1987. Nutrition of the equine athlete. Pp. 650–665 in Equine Exercise Physiology II, J. R. Gillespie and W. E. Robinson, eds. Davis, Calif.: ICEEP Publ.

Meyer, H., and L. Ahlswede. 1976. The intrauterine growth and body composition of foals and the nutrient requirements of pregnant mares. Anim. Res. Dev. 8:86 .

Meyer, H., and L. Ahlswede. 1977. Untersuchungen zum Mg-Stoffwechsel des Pferdes. Zentrabl. Veterinacrmed. 24:128.

Meyer, H., M. Schmidt, A. Lindner, and M. Pferdekamp. 1984. Beitrage zur Verdauungsphysiologie des Pferdes. 9. Einfluss einer marginalen Na-Versorgung auf Na-Bilanz. Na-Gehalt im Schwiss sowie klinische Symptome. Z. Tierphysiol. Tierernähr. Futtermittelkd. 51:182.

National Research Council. 1974. Nutrients and Toxic Substances in Water for Livestock and Poultry. Washington, D.C.: National Academy Press.

National Research Council. 1978. Nutrient Requirements of Horses, 4th rev. ed. Washington, D.C.: National Academy Press.

National Research Council. 1980. Mineral Tolerance of Domestic Animals. Washington, D.C.: National Academy Press.

National Research Council. 1982. United States-Canadian Tables of Feed Composition. Washington, D.C.: National Academy Press.

Parker, M. I. 1984. Effect of sodium chloride in feed preference and on control of feed and water intake in ponies. M.S. thesis, Cornell University.

Schryver, H. F., P. H. Craig, and H. F. Hintz. 1970. Calcium metabolism in ponies fed varying levels of calcium. J. Nutr. 100:955.

Schryver, H. F., H. F. Hintz, and P. H. Craig. 1971a. Calcium metabolism in ponies fed high phosphorus diet. J. Nutr. 101:259.

Schryver, H. F., H. F. Hintz, and P. H. Craig. 1971b. Phosphorus metabolism in ponies fed varying levels of phosphorus. J. Nutr. 101:1257.

Schryver, H. F., H. F. Hintz, J. E. Lowe, R. L. Hintz, R. B. Harper,and J. T. Reid. 1974. Mineral composition of the whole body, liver, and bone of young horses. J. Nutr. 104:126.

Schryver, H. F., O. T. Oftedal, J. Williams, L. V. Soderholm, and H. F. Hintz. 1986. Lactation in the horse: The mineral composition of mare's milk. J. Nutr. 116:2142.

Schryver, H. F., M. T. Parker, P. D. Daniluk, K. I. Pagan, J. Williams, L. V. Soderholm, and H. F. Hintz. 1987. Salt consumption and the effect of salt on mineral metabolism in horses. Cornell Vet. 77:122.

Stowe, H. D. 1971. Effects of potassium in a purified diet. J. Nutr. 101:629.

Strickland, K., F. Smith, M. Woods, and J. Jason. 1987. Dietary molybdenum as a putative copper antagonist in the horses. Eq. Vet. J. 19:50.

Swartzmann, J. A., H. F. Hintz, and H. F. Schryver. 1978. Inhibition of calcium absorption in ponies fed diets containing oxalic acid. Am. J. Vet. Res. 3:1621.

Tasker, J. B. 1980. Fluids, electrolytes, and acid-base balance. P. 425 in Clinical Biochemistry of Domestic Animals, 3rd ed., J. J. Kaneko, ed. New York: Academic Press.

Whitlock, R. H., H. F. Schryver, L. Krook, N. F. Hintz, and P. H. Craig. 1970. The effects of high dietary calcium for horses. P. 127 in Proc. 16th Am. Assoc. Eq. Pract., F. J. Milne, ed. Lexington, Ky.: Am. Assoc. Eq. Pract.

Wolter, R., J. P. Valette, and J. M. Marion. 1986. Magnesium et effort d'endurance chez le poney. Ann. Zootech. 35:255.

Trace Minerals

Arnbjerg, J. 1981. Poisoning in animals due to oral application of iron with description of a case in a horse. Nord. Veterinaermed. 33:71

Baker, H. J., and J. R. Lindsey. 1968. Equine goiter due to excess dietary iodine. J. Am. Vet. Med. Assoc. 153:1618.

Blackmore, D. J., C. Campbell, D. Cant, J. E. Holden, and J. E. Kent. 1982. Selenium status of Thoroughbreds in the United Kingdom. Eq. Vet. J. 14:139.

Bridges, C. H., J. E. Womack, E. D. Harris, and W. L. Scrutchfield. 1984. Considerations of copper metabolism in osteochondrosis of suckling foals. J. Am. Vet. Med. Assoc. 185:173.

Brown-Grant, K. 1957. The iodine concentrating mechanism of the mammary gland. J. Physiol. 135:644.

Caple, I. W., S. J. A. Edwards, W. M. Forsyth, P. Whiteley, R. H. Selth, and L. J. Fulton. 1978. Blood glutathione peroxidase activity in horses in relation to muscular dystrophy and selenium nutrition. Aust. Vet. J. 54:57.

Carbery, J. T. 1984. Osteodysgenesis in a foal associated with copper deficiency. N. Z. Vet. J. 26:280.

Coger, L. S., H. F. Hintz, H. F. Schryver, and J. E. Lowe. 1987. The effect of high zinc intake on copper metabolism and bone development in growing horses. P. 173 in Proc. 10th Eq. Nutr. Physiol. Soc. Symp. Fort Collins, Colo.: Colorado State University.

Cowgill, U. M., S. J. States, and J. E. Marburger. 1980. Smelter smoke syndrome in farm animals and manganese deficiency in northern Oklahoma, USA. Environ. Pollut. (Ser. A) 22:259.

Cromwell, G. L., V. W. Hays, and T. L. Clark. 1978. Effects of copper sulfate, copper sulfide, and sodium sulfide on performance and copper liver stores of pigs. J. Anim. Sci. 46:692.

Cromwell, G. L., T. S. Stahly, and H. J. Monegue. 1984. Effects of level and source of copper (sulfate vs. oxide) on performance and liver copper levels of weanling pigs. J. Anim. Sci. 59(Suppl.1):267.

Cupps, P. T., and C. E. Howell. 1949. The effects of feeding supplemental copper to growing foals. J. Anim. Sci. 8:286.

Cymbaluk, N. F., H. F. Schryver, and H. F. Hintz. 1981a. Copper metabolism and requirement in mature ponies. J. Nutr. 111:87.

Cymbaluk, N. F., H. F. Schryver, H. F. Hintz, D. F. Smith, and J. E. Lowe. 1981b. Influence of dietary molybdenum on copper metabolism in ponies. J. Nutr. 111:96.

Davies, M. E. 1971. The production of B$_{12}$ in the horse. Br. Vet. J. 127:34.

Dill, S. G., and W. C. Rebhun. 1985. White muscle disease in foals. Comp. Cont. Ed. 7:S627.

Drepper, K., J. O. Gutte, H. Meyer, and F. J. Schwarz. 1982. Energie- und Nahrstoffbedarf landwirtschaftlicher Nutztiere. Nr. 2 Empfehlungen zur Energie- und Nahrstoffversorgung der Pferde. Frankfurt am Main, Germany: DLG Verlag.

Drew, B., W. P. Barber, and D. G. Williams. 1975. The effect of excess dietary iodine on pregnant mares and foals. Vet. Rec. 97:93.

Driscoll, J., H. F. Hintz, and H. F. Schryver. 1978. Goiter in foals caused by excessive iodine. J. Am. Vet. Med. Assoc. 173:858.

Eamens, G. J., J. F. Macadam, and E. A. Laing. 1984. Skeletal abnormalities in young horses associated with zinc toxicity and hypocuprosis. Aust. Vet. J. 61:205.

Fadok, V. A., and S. Wild. 1983. Suspected cutaneous iodinism in a horse. J. Am. Vet. Med. Assoc. 183:1104.

Filmer, J. F. 1933. Enzootic marasmus of cattle and sheep. Aust. Vet. J. 9:163.

Food and Drug Administration. 1987. Food additives permitted in feed and drinking water of animals. Fed. Reg. 52(Part 573, No. 65):10887. April 6.

Gallagher, K., and H. D. Stowe. 1980. Influence of exercise on serum selenium and peroxide reduction system of racing Standardbreds. Am. J. Vet. Res. 41:1333.

Graham, R., J. Sampson, and H. R. Hester. 1940. The results of feeding zinc to pregnant mares and foals. Vet. Rec. 97:93.

Gunson, D. E., D. F. Kowalczyk, C. R. Shoop, and C. F. Ramberg. 1982. Environmental zinc and cadmium pollution associated with generalized osteochondrosis, osteoporosis, and nephrocalcinosis. J. Am. Vet. Med. Assoc. 180:295.

Harrington, D. D., J. Walsh, and V. White. 1973. Clinical and pathological findings in horses fed zinc deficient diets. P. 51 in Proc. 3rd Eq. Nutr. Physiol. Soc. Symp. Gainesville, Fla.: University of Florida.

Howes, A. D., and I. A. Dyer. 1971. Diet and supplemental mineral effects on manganese metabolism in newborn calves. J. Anim. Sci. 32:141.

Jarrige, R., and W. Martin-Rosset. 1981. Le cheval: Reproduction, selection, alimentation, exploitation. XIII Journées du Grenier de Theix. Paris: Institut National de la Recherche Agronomique.

Jiong, Z., C. Zeng-Cheng, L. Su-Mei, D. Yin Jie, Z. Kang-Nan, and Z. Xu-Jiu. 1987. Selenium deficiency in horses (1981–1983). P. 843 in Selenium in Biology and Medicine, Part B, G. F. Combs, J. E. Spallholz, O. A. Levander, and J. E. Oldfield, eds. New York: Van Nostrand Reinhold.

Kirkham, W. W., H. Guttridge, J. Bowden, and G. T. Edds. 1971. Hematopoietic responses to hematinics in horses. J. Am. Vet. Med. Assoc. 159:1316.

Knight, D. A., A. A. Gabel, S. M. Reed, L. R. Bramlage, W. J. Tyznik, and R. M. Embertson. 1985. Correlation of dietary mineral to incidence and severity of metabolic bone disease in Ohio and Kentucky. P. 445 in Proc. 31st Am. Assoc. Eq. Pract., F. J. Milne, ed. Lexington, Ky.: Am. Assoc. Eq. Pract.

Knight, D. A., S. E. Weisbrade, L. M. Schmall, and A. A. Gavel. 1988. Copper supplementation and cartilage lesions in foals. Pp. 191–194 in Proc. 33rd Annu. Conv. Am. Assoc. Eq. Pract. New Orleans, La.

Kruzhova, E. 1968. Mikroelementy i vosproizvoditel'naja funkeija kobyl. Tr. Vses. Inst. Konevodstvo. 2:28 (as cited in Nutr. Abstr. Rev. 39:807, 1968).

Lawrence, L. 1986. The use of non-invasive techniques to estimate bone mineral content and bone strength in the horse. Ph.D. dissertation, University of Florida.

Lawrence, L. A., E. A. Ott, R. L. Asquith, and G. J. Miller. 1987. Influence of dietary iron on growth, tissue mineral composition, apparent phosphorus absorption, and chemical properties of bone. Pp. 563 in Proc. 10th Eq. Nutr. Physiol. Soc. Symp. Fort Collins, Colo.: Colorado State University.

Lindholm, A., and A. Asheim. 1971. Vitamin E and certain muscular enzymes in the blood serum of horses. Acta Agr. Scand. (Suppl. 19).

Maylin, G. H., D. S. Rubin, and D. H. Lein. 1980. Selenium and vitamin E in horses. Cornell Vet. 70:272.

Messer, N. T. 1981. Tibiotarsal effusion associated with chronic zinc intoxication in three horses. J. Am. Vet. Med. Assoc. 178:294.

Meyer, H. 1986. Mineral requirements of riding horses. Paper presented at IV World Congress of Animal Feeding, Madrid, June 30–July 4.

Miller, W. T., and K. T. Williams. 1940. Minimal lethal dose of selenium as sodium selenite in horses, mules, cattle, and swine. J. Agr. Res. 60:163.

Moore, C. V. 1951. Iron metabolism and nutrition. Harvey Lect. 55:67

Mullaney, T., and C. Brown. 1988. Iron toxicity in neonatal foals. Eq. Vet. J. 20:119.

National Research Council. 1978. Nutrient Requirements of Horses. Washington, D.C.: National Academy Press.

National Research Council. 1980. Mineral Tolerance of Domestic Animals. Washington, D.C: National Academy Press.

Paglia, D. E., and W. H. Valentine. 1967. Studies on the quantitative and qualitative characterization of erythrocyte glutathione peroxidase. J. Lab. Clin. Med. 70:158.

Rodenwold, B. W., and B. T. Simms. 1935. Iodine for brood mares. Proc. Am. Soc. Anim. Prod. 34:89.

Rojas, M. A., I. A. Dyer, and W. A. Cassatt. 1965. Manganese deficiency in bovine. J. Anim. Sci. 24:664.

Roneus, B. O., and B. Lindholm. 1983. Glutathione peroxidase activity in blood of healthy horses given different selenium supplementation. Nord. Veterinaermed. 35:337.

Rosenfeld, I., and O. A. Beath. 1964. Selenium: Geobotany, Biochemistry, Toxicity, and Nutrition. New York: Academic Press.

Rotruck, J. T., A. L. Pope, H. E. Ganther, A. M. Swanson, D. G. Hafeman, and W. G. Hoekstra. 1973. Selenium: Biochemical role as a component of glutathione peroxidase. Science 179:588.

Salminen, K. 1975. Cobalt metabolism in horses: Serum level and biosynthesis of vitamin B$_{12}$. Acta Vet. Scand. 16:84.

Schryver, H. F., H. F. Hintz, J. E. Lowe, R. L. Hintz, R. B. Harper, and J. T. Reid. 1974. Mineral composition of the whole body, liver, and bone of young horses. J. Nutr. 104:126.

Shellow, J. S., S. G. Jackson, J. P. Baker, and A. H. Cantor. 1985. The influence of dietary selenium levels on blood levels of selenium and glutathione peroxidase activity in the horse. J. Anim. Sci. 61:590.

Shupe, J. L., and A. E. Olson. 1971. Clinical aspects of fluorosis in horses. J. Am. Vet. Med. Assoc. 158:167.

Smith, J. D., R. M. Jordan, and M. L. Nelson. 1975. Tolerance of ponies to high levels of dietary copper. J. Anim. Sci. 41:1645.

Smith, J. E., K. Moore, J. E. Cipriano, and P. G. Morris. 1984. Serum ferritin as a measure of stored iron in horses. J. Nutr. 114:677.

Stowe, H. D. 1967. Serum selenium and related parameters of naturally and experimentally fed horses. J. Nutr. 93:60.

Stowe, H. D. 1968. Effects of age and impending parturition upon serum copper of Thoroughbred mares. J. Nutr. 95:179.

Stowe, H. D. 1980. Effects of copper pretreatment upon the toxicity of selenium in ponies. Am. J. Vet. Res. 41:1925.

Strickland, K., F. Smith, M. Woods, and J. Mason. 1987. Dietary molybdenum as a putative copper antagonist in the horse. Eq. Vet. J. 19:50

Traub-Dargatz, J. L., and D. W. Hamar. 1986. Selenium toxicity in horses. Comp. Cont. Vet. Ed. (Eq.) 8:771.

Ullrey, D. E., W. T. Ely, and R. L. Covert. 1974. Iron, zinc, and copper in mare's milk. J. Anim. Sci. 38:1276

Underwood, E. J. 1977. Trace elements in human and animal nutrition. New York: Academic Press.

Underwood, E. J. 1981. The Mineral Nutrition of Livestock, 2nd ed. Slough, England: Commonwealth Agricultural Bureaux.

Whetter, P. A., and D. E. Ullrey. 1978. Improved fluorometric method for determining selenium. J. Assoc. Off. Anal. Chem. 61:927.

Wright, P. L., and M. C. Bell. 1966. Comparative metabolism of selenium and tellurium in sheep and swine. Am. J. Physiol. 211:6.

Young, J. K., G. D. Potter, L. W. Greene, S. P. Webb, J. W. Evans, and G. W. Webb. 1987. Copper balance in miniature horses fed varying amounts of zinc. P. 173 in Proc. 10th Eq. Nutr. Physiol. Soc. Symp. Fort Collins, Colo.: Colorado State University.

VITAMINS

Fat-Soluble Vitamins

Ahlswede, L., and H. Konermann. 1980. Erfahrungen mit der Oralen und Parenteralen Applikation von Beta-Carotin beim Pferd. Prakt. Tierarzt 61:47.

Alam, S. Q., and B. S. Alam. 1981. Effects of excess vitamin E on rat teeth. Calcif. Tissue Int. 33:619.

American Academy of Pediatrics, Committee on Nutrition. 1971. Vitamin K supplementation for infants receiving milk substitute infant formulas and for those with fat malabsorption. Pediatrics 48:483.

Baalsrud, K. J., and G. Overnes. 1986. The influence of vitamin E and selenium supplements on antibody production in horses. Eq. Vet. J. 18:472.

Bendich, A., E. Gabriel, and L. J. Machlin. 1986. Dietary vitamin E requirement for optimum immune responses in the rat. J. Nutr. 116:675.

Bille, N. 1970. Hypervitaminosis D og calciphylaxis hos husdyr. Nord. Veterinaermed. 22:218.

Butler, P., and D. J. Blackmore. 1983. Vitamin E values in the plasma of stabled Thoroughbred horses in training. Vet. Rec. 112:60.

Corrigan, J. J., Jr. 1979. Coagulation problems relating to vitamin E. Am. J. Pediat. Hematol. Oncol. 1:169.

Davies, K. J. A., A. T. Quintanilha, G. A. Brooks, and L. Packer. 1982. Free radicals and tissue damage produced by exercise. Biochem. Biophys. Res. Commun. 107:1198.

Dobereiner, J., C. H. Tokarnia, J. B. D. Da Costa, J. L. E. Campos, and M. S. Dayrell. 1971. "Espichamento," intoxicaco de bovinos por Solanum malacoxylon, no pantanal de moto gross. Presq. Agropec. Bras. Ser. Vet. 6:91.

Donoghue, S., D. S. Kronfeld, S. J. Berkowitz, and R. L. Copp. 1981. Vitamin A nutrition of the equine: Growth, serum biochemistry, and hematology. J. Nutr. 111:365.

Edwards, J. T. 1937. Carotene (or vitamin A) deficiency and some common unsoundnesses in horses: A comparative review of their possible relationship. J. R. Army Vet. Corp. 9(3):60.

Eitzer, P., and H. J. Rapp. 1985. Zur Oralen Anwendung von Synthetischen Beta-Carotin bei Zuchstuten. Prakt. Tierarzt 66:123.

El Shorafa, W. M., J. P. Feaster, E. A. Ott, and R. L. Asquith. 1979. Effect of vitamin D and sunlight on growth and bone development of young ponies. J. Anim. Sci. 48:882.

Ferraro, J., and J. F. Cote. 1984. Broodmare management techniques improve conception rates. Standardbred 12:56.

Fonnesbeck, P. V., and L. D. Symons. 1967. Utilization of carotene of hay by horses. J. Anim. Sci. 26:1030.

Garton, C. L., G. W. VanderNoot, and P. V. Fonnesbeck. 1964. Seasonal variation in carotene and vitamin A concentrations of the blood of brood mares in New Jersey. J. Anim. Sci. 23:1233 (Abstract).

Guilbert, H. R., C. E. Howell, and G. H. Hart. 1940. Minimum vitamin A and carotene requirements of mammalian species. J. Nutr. 19:91.

Harrington, D. D. 1982. Acute vitamin D$_2$ (ergocalciferol) toxicosis in horses: Case report and experimental studies. J. Am. Vet. Med. Assoc. 180:867.

Harrington, D. D., and E. H. Page. 1983. Acute vitamin D$_3$ toxicosis in horses: Case reports and experimental studies of the comparative toxicity of vitamin D$_2$ and D$_3$. J. Am. Vet. Med. Assoc. 182:1358.

Hintz, H. F., H. F. Schryver, J. E. Lowe, J. King, and L. Krook. 1973. Effect of vitamin D on Ca and P metabolism in ponies. J. Anim. Sci. 37:282.

Horst, R. L., T. A. Reinhardt, J. R. Russell, and J. L. Wapoli. 1984.

Isolation and identification of vitamin D_2 and vitamin D_3 from *Medicago sativa* (alfalfa plant). Arch. Biochem. Biophys. 231:67.

Howell, C. E., G. H. Harts, and N. R. Ittner. 1941. Vitamin A deficiency in horses. Am. J. Vet. Res. 2:60.

Hughes, M. R., T. A. McCain, S. Y. Chang, and M. R. Haussler. 1977. Presence of 1,25-dihydroxy vitamin D_3-glycoside in the calcinogenic plant *Cestrum diurnum*. Nature 268:347.

Jackson, M. J., D. A. Jones, and R. H. T. Edwards. 1983. Vitamin E and skeletal muscle. Pp. 224–239 in Biology of Vitamin E, Ciba Foundation Symposium 101. London: Pitman Books.

Jarrett, S. H., W. A. Schurg, and B. L. Reid. 1987. Plasma fractions of vitamin A alcohol, palmitate, and acetate in horses fed deficient and excess dietary vitamin A. Pp. 1–4 in Proc. 10th Eq. Nutr. Physiol. Soc. Symp. Fort Collins, Colo.: Colorado State University.

Krook, L. P., R. H. Wasserman, J. N. Shively, A. H. Tashjian, Jr., T. D. Brokken, and J. F. Morton. 1975. Hypercalcemia and calcinosis in Florida horses: Implication of the shrub, *Cestrum diurnum*, as the causative agent. Cornell Vet. 65:26.

Liu, S. K., E. P. Dolensek, C. R. Adams, and J. P. Tappe. 1983. Myelopathy and vitamin E deficiency in six Mongolian wild horses. J. Am. Vet. Med. Assoc. 183:1266.

Mäenpää, P. H., T. Koskinen, and E. Koskinen. 1988a. Serum profiles of vitamins A, E, and D in mares and foals during different seasons. J. Anim. Sci. 66:1418.

Mäenpää, P. H., A. Pirhonen, and E. Koskinen. 1988b. Vitamin A, E, and D nutrition in mares and foals during the winter season: Effect of feeding two different vitamin-mineral concentrates. J. Anim. Sci. 66:1424.

March, B. E., E. Wong, L. Seier, J. Sim, and J. Biely. 1973. Hypervitaminosis E in the chick. J. Nutr. 103:371.

Mayhew, I. G., C. M. Brown, H. D. Stowe, A. L. Trapp. F. J. Derksen, and S. F. Clement. 1987. Equine degenerative myeloencephalopathy: A vitamin E deficiency that may be familial. J. Vet. Intern. Med. 1:45.

Meigs, E. B. 1939. The feeding of dairy cows for intensive milk production in practice. Pp. 566–596 in U.S.D.A. Yearbook. Washington, D.C.: Department of Agriculture.

Molitor, H., and J. Robinson. 1940. Oral and parenteral toxicity of vitamin K_1, phthiocol and 2-methyl 1,4-naphthoquinone. Proc. Soc. Exp. Biol. Med. 43:125.

National Research Council. 1987. Vitamin Tolerance of Animals. Washington, D.C.: National Academy Press.

Nieberle, R., and S. Chors. 1954. Lehrbuch der Speziellen Pathologischer Anatomia der Holasterire. Jena, E. Germany: Gustav Fischer.

Park, E. A. 1923. The etiology of rickets. Physiol. Rev. 3:106.

Quintanilha, A. T., and L. Packer. 1983. Vitamin E, physical exercise, and tissue oxidative damage. Pp. 56–69 in Biology of Vitamin E, Ciba Foundation Symposium 101. London: Pitman Books.

Ralston, S. L., S. A. Jackson, V. A. Rich, and E. L. Squires. 1985. Effect of vitamin A supplementation on the seminal characteristics and sexual behavior of stallions. P. 74 in Proc. 9th Eq. Nutr. Physiol. Soc. Symp. East Lansing, Mich.: Michigan State University.

Rebhun, W. C., B. C. Tennant, S. G. Dill, and J. M. King. 1984. Vitamin K_3-induced renal toxicosis in the horse. J. Am. Vet. Med. Assoc. 184:1237.

Roneus, B. O., R. V. J. Hakkarainen, C. A. Lindholm, and J. I. Tyopponen. 1986. Vitamin E requirements of adult Standardbred horses evaluated by tissue depletion and repletion. Eq. Vet. J. 18:50.

Schougaard, H., A. Basse, G. Gissel-Nielsen, and M. G. Simesen. 1972. Nutritional muscular dystrophy (NMD) in foals. Nord. Veterinaermed. 24:67.

Stowe, H. D. 1968a. Alpha-tocopherol requirements for equine erythrocyte stability. Am. J. Clin. Nutr. 21:135.

Stowe, H. D. 1968b. Experimental equine avitaminosis A and E. P. 27

in Proc. 1st Eq. Nutr. Res. Soc. Lexington, Ky.: University of Kentucky.

Van der Holst, M. 1984. Experiences with oral administration of beta-carotene to pony mares in early spring. P. 6 in Proc. 35th Annu. Meet. Eur. Assoc. Anim. Prod.

Vermeer, C., and M. Ulrich. 1982. Vitamin K-dependent carboxylase in horse liver, spleen, and kidney. Thromb. Res. 28:171.

Waite, R., and K. N. S. Sastry. 1949. The carotene content of dried grass. J. Agr. Sci. 39:174

Wasserman, R. H., J. D. Henion, M. R. Haussler, and T. A. McCain. 1976. Calciogenic factor in *Solanum malacoxylon*: Evidence that it is 1,25-dihydroxyvitamin D_3-glycoside. Science 194:853.

Wilson, T. M., H. A. Morrison, N. C. Palmer, G. G. Finley, and A. A. van Dreumel. 1976. Myodegeneration and suspected selenium/vitamin E deficiency in horses. J. Am. Vet. Med. Assoc. 169:213.

Worker, N. A., and B. J. Carrillo. 1967. "Entique Seco," calcification and wasting in grazing animals in the Argentine. Nature 215:72.

Yang, N. Y. J., and I. D. Desai. 1977a . Effect of high levels of dietary vitamin E on hematological indices and biochemical parameters in rats. J. Nutr. 107:1410.

Yang, N. Y. J., and I. D. Desai. 1977b. Effect of high levels of dietary vitamin E on liver and plasma lipids and fat soluble vitamins in rats. J. Nutr. 107:1418.

Water-Soluble Vitamins

Alexander, F., and M. E. Davies. 1969. Studies on vitamin B_{12} in the horse. Br. Vet. J. 125:169.

Allen, B. V. 1978. Serum folate levels in horses with particular reference to the English Thoroughbred. Vet. Rec. 103:257.

Allen, B. V. 1984. Dietary intake and absorption of folic acid in the horse. Proc. Assoc. Vet. Clin. Pharmacol. Ther. 8:118.

Anderson, P. A., D. H. Baker, and S. P. Mistry. 1978. Bioassay determination of the biotin content of corn, barley, sorghum, and wheat. J. Anim. Sci. 47:654.

Bauchop, T., and L. King. 1968. Amprolium and thiamine pyrophosphotransferase. Appl. Microbiol. 16:961.

Bertone, J. J., H. F. Hintz, and H. F. Schryver. 1984. Effect of caffeic acid on thiamin status in ponies. Nutr. Rep. Int. 30:281.

Caple, I. W., G. G. Halpin, J. K. Azuolas, G. F. Nugent, and R. J. Cram. 1982. Studies of selenium, iodine, and vitamin B_{12} nutrition of horses in Victoria. Pp. 57–68 in Proc. 4th Bain-Fall Memorial Lecture. Sydney, Australia: Sydney University.

Carlstrom, B., and A. Hjarre. 1939. Durch B-avitaminose verursachte Mangelkrankheiten bei Militärpferden. Tierernaehrung 11:121.

Carroll, F. D. 1950. B vitamin content in the skeletal muscle of the horse fed a B vitamin low diet. J. Anim. Sci. 9:139.

Carroll, F. D., H. Goss, and C. E. Howell. 1949. The synthesis of B vitamins in the horse. J. Anim. Sci. 8:290.

Cello, R. M. 1962. Recent findings in periodic ophthalmia. P. 39 in Proc. 8th Annu. Assoc. Eq. Pract. San Francisco.

Comben, N., R. J. Clark, and D. J. B. Sutherland. 1984. Clinical observations on the response of equine hoof defects to dietary supplementation with biotin. Vet. Rec. 115:642.

Cunha, T. J., D. C. Lindley, and M. E. Ensminger. 1946. Biotin deficiency syndrome in pigs fed desiccated egg white. J. Anim. Sci. 5:219.

Cunha, T. J., R. W. Colby, L. K. Bustad, and J. F. Bone. 1948. The need for and interrelationship of folic acid, anti-pernicious anemia liver extract, and biotin in the pig. J. Nutr. 36:215.

Cymbaluk, N. F., P. B. Fretz, and F. M. Loew. 1978. Amprolium-induced thiamine deficiency in horses: Clinical features. Am. J. Vet. Res. 39:255.

Davies, M. E. 1971. The production of B_{12} in the horse. Br. Vet. J. 127:34.

Davis, G. K., and C. L. Cole. 1943. The relation of ascorbic acid to breeding performance in horses. J. Anim. Sci. 2:53.

Diniz, J. M. F., J. R. Bashe, and N. J. de Camargo. 1984. Intoxicacao natural de asininos por *Pteridium aquilinum* (L.) Kühn no Brazil. Arquivo Brasileiro de Med. Vet. Zootech. 36:512.

Ender, F., and A. Helgebostad. 1972. Iron deficiency anemia in mink. Z. Tierphysiol. Tierernaehr. Futtermittelkd. 19:22.

Errington, B. J., B. S. Hodgkiss, and E. P. Jayne. 1954. Ascorbic acid in certain body fluids of horses. Am. J. Vet. Res. 3:242.

Filmer, J. F. 1933. Enzootic marasmus of cattle and sheep. Aust. Vet. J. 9:163.

Haley, T. J., and A. M. Flesher. 1946. A toxicity study of thiamine hydrochloride. Science 104:567.

Hawkins, D. R. 1968. Treatment of schizophrenia based on the medical model. J. Schiz. 2:3.

Heimann, W., K. Hermann, and G. Feucht. 1971. Über das vorkommen der Hydroxyzimtsauren in Gemüse. Z. Lebensm. Forsch. 145:20, 199, 206.

Irvine, C. H. G., and N. G. Prentice. 1962. The effect of large doses of thiamine on the horse. N. Z. Vet. J. 10:86.

Jaeschke, G., and H. Keller. 1982. The ascorbic acid status of horses. 4. Behavior of intravenously applied ascorbic acid in the serum. Berl. Muench. Tieraerztl. Wochenschr. 95:71.

Jones, T. C. 1942. Equine periodic ophthalmia. Am. J. Vet. Res. 3:45.

Jones, T. C., F. D. Maurer, and T. O. Roby. 1945. The role of nutrition in equine periodic ophthalmia. Am. J. Vet. Res. 6:67.

Jones, T. C., T. O. Roby, and F. D. Maurer. 1946. The relation of riboflavin to equine periodic ophthalmia. Am. J. Vet. Res. 7:403.

Jordan, R. M. 1979. Effect of thiamin and vitamin A and D supplementation on growth of weanling ponies. Pp. 67–68 in Proc. 6th Eq. Nutr. Physiol. Soc. Symp. College Station, Tex.: Texas A&M University.

Kempson, S. A. 1987. Scanning electron microscope observations of hoof horn from horses with brittle feet. Vet. Rec. 120:568.

Konishi, T., and S. Ichijo. 1984. Experimentally induced equine bracken poisoning by thermostabile antithiamine factor (SF factor) extracted from dried bracken. J. Jpn. Vet. Med. Assoc. 37:730.

Krinke, G., H. H. Schaumberg, P. S. Spencer, J. Suter, P. Thomann, and R. Hess. 1980. Pyridoxine megavitaminosis produces degeneration of peripheral sensory neurons (sensory neuronopathy) in the dog. Neurotoxicology 2:13.

Linerode, P. A. 1966. Studies on the synthesis and absorption of B-complex vitamins in the equine. Ph.D. dissertation, Ohio State University, Columbus.

Linerode, P. A. 1967. Studies on the synthesis and absorption of B-complex vitamins in the horse. Am. Assoc. Eq. Pract. 13:283.

Loscher, W., G. Jaeschke, and H. Keller. 1984. Pharmacokinetics of ascorbic acid in horses. Eq. Vet. J. 16:59.

Lott, D. G. 1951. The use of thiamine in mare's tail poisoning of horses. Can. J. Comp. Med. Vet. Sci. 15:274.

Mackay, A. 1961. Some effects of drugs "in the doping of" racehorses. N. Z. Vet. J. 9:129.

Martin, A. A. 1975. Nigro pallidal encephalomalacia in horses caused by chronic poisoning with yellow star thistle. Nutr. Abstr. Rev. 45:85.

Menon, I. A., and E. Sognen. 1971. Amprolium and transport of thiamine in suspensions of intestinal cells. Acta. Vet. Scand. 12:111.

Naito, K., T. Shimanura, and K. Kuwabara. 1925. Experimental studies on the effect of polished rice feeding and its B defect in the horse. 3rd Rep. Jpn. Gov. Inst. Vet. Res. 3:51.

National Research Council. 1982. United States-Canadian Tables of Feed Composition, 3rd. rev. Washington, D.C.: National Academy Press.

National Research Council. 1987. Vitamin Tolerance of Animals. Washington, D.C.: National Academy Press.

National Research Council. 1988. Nutrient Requirements of Swine. Washington, D.C.: National Academy Press.

Pearson, P. B., and R. W. Luecke. 1944. Studies on the metabolism of nicotinic acid in the horse. Arch. Biochem. 6:63.

Pearson, P. B., and H. Schmidt. 1948. Pantothenic acid studies with the horse. J. Anim. Sci. 7:78.

Pearson, P. B., M. K. Sheybani, and H. Schmidt. 1943. The metabolism of ascorbic acid in the horse. J. Anim. Sci. 2:175.

Pearson, P. B., M. K. Sheybani, and H. Schmidt. 1944a. The B vitamin requirements of the horse. J. Anim. Sci. 3:166.

Pearson, P. B., M. K. Sheybani, and H. Schmidt. 1944b. Riboflavin in the nutrition of the horse. Arch. Biochem. 6:467.

Roberts, H. E., E. T. Evans, and W. C. Evans. 1949. The production of bracken staggers in the horse and its treatment by B_1 therapy. Vet. Rec. 61:549.

Roberts, M. C. 1983. Serum and red cell folate and serum vitamin B_{12} levels in horses. Aust. Vet. J. 60:106.

Roberts, S. J. 1958. Sequelae of leptospirosis in horses on a small farm. J. Am. Vet. Med. Assoc. 133:189.

Robie, T. R. 1967. Cyproheptadine: An excellent antidote for niacin-induced hyperthermia. J. Schiz. 1:133.

Salminen, K. 1975. Cobalt metabolism in horses: Serum level and biosynthesis of vitamin B_{12}. Acta Vet. Scand. 16:84.

Schaller, K., and H. Holler. 1976. Thiamine absorption in the rat. IV. Effects of caffeic acid (3,4-dihydroxycinnamic acid) upon absorption and active transport of thiamine. Int. J. Vit. Nutr. Res. 46:143.

Schaumberg, H., J. Kaplan, A. Windebank, N. Vick, S. Rasmus, D. Pleasure, and M. J. Brown. 1983. Sensory neuropathy from pyridoxine abuse: A new megavitamin syndrome. N. Engl. J. Med. 309:445.

Schumacher, M. F., M. A. Williams, and R. L. Lyman. 1965. Effect of high intakes of thiamine, riboflavin, and pyridoxine on reproduction in rats and vitamin requirements of the offspring. J. Nutr. 86:343.

Schweigert, B. S., P. B. Pearson, and M. C. Wilkening. 1947. The metabolic conversion of tryptophan to nicotinic acid and to N^1-methyl nicotinamide. Arch. Biochem. 12:139.

Seckington, I. M., R. G. Hunstman, and G. C. Jenkins. 1967. The serum folic acid levels of grass-fed and stabled horses. Vet. Rec. 81:158.

Snow, D. H., and M. Frigg. 1987a. Plasma concentrations at monthly intervals of ascorbic acid, retinol, β-carotene, and α-tocopherol in two Thoroughbred racing stable and effects of supplementation. Pp. 55–60 in Proc. 10th Eq. Nutr. Physiol. Soc. Symp. Fort Collins, Colo.: Colorado State University.

Snow, D. H., and M. Frigg. 1987b. Oral administration of different formulations of ascorbic acid to the horse. Pp. 617–624 in Proc. 10th Eq. Nutr. Physiol. Soc. Symp. Fort Collins, Colo.: Colorado State University.

Stewart, G. A. 1972. Drugs, performance, and response to exercise in the racehorse. 2. Observations on amphetamine, promazine, and thiamine. Aust. Vet. J. 48:544.

Stillions, M. C., S. M. Teeter, and W. E. Nelson. 1971a. Ascorbic acid requirement of mature horses. J. Anim. Sci. 32:249.

Stillions, M. C., S. M. Teeter, and W. E. Nelson. 1971b. Utilization of dietary B_{12} and cobalt by mature horses. J. Anim. Sci. 32:252.

Topliff, D. R., G. D. Potter, J. L. Kreider, and C. R. Creagor. 1981. Thiamin supplementation for exercising horses. Pp. 167–172 in Proc. 7th Eq. Nutr. Physiol. Soc. Symp. Warrenton, Va.

Unna, K., and J. S. Greslin. 1941. Studies of the toxicity and pharmacology of pantothenic acid. J. Pharmacol. Exp. Ther. 73:85.

Winter, C. A., and C. W. Mushett. 1951. Absence of toxic effects from single injections of crystalline vitamin B$_{12}$. J. Am. Pharm. Assoc. 39:360.

Winter, S. L., and J. L. Boyer. 1973. Hepatic toxicity from large doses of vitamin B-3 nicotinamide. N. Engl. J. Med. 289:1180.

WATER

Argenzio, R. A., J. E. Lowe, D. W. Pickard, and C. E. Stevens. 1974. Digesta passage and water exchange in the equine large intestine. Am. J. Physiol. 226:1035.

Caljuk, E. A. 1961. Water metabolism and water requirements of horses. Tr. Vses. Inst. Konevodstvo. 23:295 (as cited in Nutr. Abstr. Rev. 32:574, 1962).

Carlson, G.P. 1982. Thermoregulation and fluid balance in the exercising horse. Pp. 291–309 in Proc. 1st Int. Conf. Eq. Exercise Physiol. Oxford.

Carlson, G. P., and R. A. Mansmann. 1974. Serum electrolyte plasma protein alterations in horses used in endurance rides. J. Am. Vet. Med. Assoc. 165:262.

Carlson, G. P., and P. O. Ocen. 1979. Composition of equine sweat following exercise in high environmental temperatures and in response to intravenous epinephrine administration. J. Eq. Med. Surg. 3:27–32.

Fonnesbeck, P. V. 1968. Consumption and excretion of water by horses receiving all hay and hay-grain diets. J. Anim. Sci. 27:1350–1356.

Frape, D. L., C. K. Peace, and P. M. Ellis. 1979. Some physiological changes in a fit and unfit horse associated with a long distance ride. Proc. 30th Annu. Meet. Eur. Assoc. Anim. Prod. (cited in Frape, D. L. ed. 1986. Equine Nutrition and Feeding. Toronto: Longman Scientific & Technical Publ.).

Hintz, H. F. 1983. Nutritional requirements of the exercising horse—A review. Pp. 275–291 in Proc. 1st Int. Conf. Eq. Exercise Physiol. Oxford.

Kerr, M. G., and D. H. Snow. 1983. The composition of equine sweat during prolonged adrenaline infusion, heat exposure and exercise. Am. J. Vet. Res. 44:1571.

Leitch, I., and J. S. Thomson. 1944. The water economy of farm animals. Nutr. Abstr. Rev. 14:197.

Lewis, L. D. 1982. Feeding and Care of the Horse. Philadelphia: Lea & Febiger.

Lucke, J. N., and G. M. Hall. 1978. Biochemical changes in horses during a 50-mile endurance ride. Vet. Rec. 102:356.

Meyer, H. V., C. Winkel, L. Ahlswede, and C. Weidenhaupt. 1978. Untersuchungen über Schweissmenge und Schweiss-zusammensetzung beim Pferd. Tierarztl Umsch 33:330 (cited in Frape, D. L., ed. 1986. Equine Nutrition and Feeding. Toronto: Longman Scientific & Technical Publ.).

Mitchell, H. H. 1962. The water requirements for maintenance. P. 701 in Comparative Nutrition of Man and Domestic Animals. New York: Academic Press.

Morrison, F. B. 1937. Feeds and Feeding. Ithaca, N.Y.: Morrison Publ. Co.

National Research Council. 1974. Nutrients and Toxic Substances in Water for Livestock and Poultry. Washington, D.C.: National Academy Press.

Puls, R. 1988. Mineral Levels in Animal Health-Diagnostic Data. British Columbia, Canada: Sherpa, International.

Robinson, J. R., and R. A. McCance. 1952. Water metabolism. Annu. Rev. Physiol. 14:115.

Rose, R. J., R. A. Purdue, and W. Hensley. 1977. Plasma biochemistry alterations during an endurance ride. Eq. Vet. J. 9:122.

Rose, R. J., K. S. Arnold, S. Church, and R. Paris. 1980. Plasma and sweat electrolyte concentrations in the horse during long distance exercise. Eq. Vet. J. 12:19.

Saltin, B. 1964. Aerobic and anaerobic work capacity after dehydration. J. Appl. Physiol. 19:1114.

Snow, D. H., M. G. Kerr, M. A. Nimmo, and E. M. Abbott. 1982. Alterations in blood, sweat, urine and muscle composition during prolonged exercise in the horse. Vet. Rec. 110:337.

White, K. K., C. E. Short, H. F. Hintz, M. W. Ross, F. R. Lesser, P. F. Leids, J. E. Lowe, and H. F. Schryver. 1978. The value of dietary fat for working horses. II. Physical evaluation. J. Eq. Med. Surg. 2:525.

PHYSICAL CHARACTERISTICS AND SUITABILITY OF FEEDS

Adams, L. G., J. W. Dollahite, W. M. Romane, T. L. Bullard, and C. H. Bridges. 1969. Cystitis and ataxia associated with sorghum ingestion by horses. J. Am. Vet. Med. Assoc. 155:518.

Aust, S. D., H. P. Broquist, and K. L. Rinehart, Jr. 1968. Slaframine: A parasympathomimetic from Rhizoctonia leguminicola. Biotech. Bioeng. 10:403.

Barnett, D. T., S. G. Jackson, and J. P. Baker. 1984. Fescue toxicity in the broodmare. Equiline. EL 1–2. In Equine Data Line, September 1984. Ky. Horsemen's Ed. Found. Coop. Ext. Serv., University of Kentucky.

Battle, G. H., S. G. Jackson, and J. P. Baker. 1988. Acceptability and digestibility of preservative-treated hay by horses. Nutr. Rep. Int. 37:83.

Beasley, V. R., G. R. Wolf, D. C. Fischer, A. C. Ray, and W. C. Edwards. 1983. Cantharidin toxicosis in horses. J. Am. Vet. Med. Assoc. 182:283.

Bowman, V. A., J. P. Fontenot, K. E. Webb, Jr., and T. N. Meacham. 1977. Digestion of fat by equine. P. 40 in Proc. 5th Eq. Nutr. Physiol. Soc. Symp. St. Louis, Mo.

Cheeke, P. R., and L. R. Shull. 1985. Natural Toxicants in Feeds and Poisonous Plants. Westport, Conn.: AVI Publ.

Christensen, C. 1987. Blister beetles and alfalfa. Pp. 10–11 in Equine Data Line, September. Kentucky Horsemen's Ed. Found. Coop. Ext. Serv., University of Kentucky.

Cunha, T. J. 1980. Horse Feeding and Nutrition. New York: Academic Press.

Cymbaluk, N. F., and D. A. Christensen. 1986. Nutrient utilization of pelleted and unpelleted forages by ponies. Can. J. Anim. Sci. 66:237.

Cymbaluk, N. F., J. D. Millar, and D. A. Christensen. 1986. Oxalate concentration in feeds and its metabolism by ponies. Can. J. Anim. Sci. 66:1107.

Darlington, J. M., and T. V. Hershberger. 1968. Effect of forage maturity on digestibility, intake, and nutritive value of alfalfa, timothy, and orchardgrass by equine. J. Anim. Sci. 27:1572.

Duren, S. E., S. G. Jackson, and J. P. Baker. 1986. Effect of dietary fat on blood parameters in exercised Thoroughbred horses. Proc. 2nd Int. Conf. Eq. Exercise Physiol. San Diego, Calif.

Ensminger, M. E., and C. G. Olentine. 1978. Feeds & Nutrition, 1st ed. Clovis, Calif.: Ensminger Publ.

Fonnesbeck, P. V. 1969. Partitioning of the nutrients of forage for horses. J. Anim. Sci. 28:624.

Fonnesbeck, P. V., and L. D. Symons. 1967. Utilization of carotene of hay by horses. J. Anim. Sci. 26:1030.

Frape, D. L. 1986. Equine Nutrition and Feeding, 1st ed. New York: Churchill Livingstone.

Garrett, L. W., E. D. Heimann, W. H. Pfander, and L. L. Wilson. 1980. Reproductive problems of pregnant mares grazing fescue pastures. J. Anim. Sci. 51(Suppl.1):237.

Haenlein, C. F. W., R. D. Holdren, and Y. M. Yoon. 1966. Comparative response of horses and sheep to different physical forms of alfalfa hay. J. Anim. Sci. 25:740.

Hambleton, P. L., L. M. Slade, D. W. Hamar, E. W. Kienholz, and L. D. Lewis. 1980. Dietary fat and exercise conditioning effect on metabolic parameters in the horse. J. Anim. Sci. 51:1330.

Harper, F., and J. Henton. 1981. Reproductive problems in pregnant broodmares associated with fescue forage. The Morgan Horse 41:78.

Hintz, H. F. 1983. Horse Nutrition—A Practical Guide. New York: Arco Publ.

Hintz, H. F., and R. G. Loy. 1966. Effects of pelleting on nutritive value of horse rations. J. Anim. Sci. 25:1059.

Hintz, H. F., M. W. Ross, F. R. Lesser, P. F. Leids, K. K. White, J. E. Lowe, C. E. Short, and H. F. Schryver. 1978. The value of dietary fat for working horses. I. Biochemical and hematological evaluations. J. Eq. Med. Surg. 2:483.

Hintz, H. F., J. E. Lowe, and W. F. Miller. 1983. Studies on the feeding of hay treated with a mixture of propionic and acetic acids to horses. P. 1 in Proc. 8th Eq. Nutr. Physiol. Soc. Symp. Lexington, Ky.: University of Kentucky.

Hintz, H. F., H. F. Schryver, J. Doty, C. Lakin, and R. A. Zimmerman. 1984. Oxalic acid content of alfalfa hays and its influence on the availability of calcium, phosphorus and magnesium to ponies. J. Anim. Sci. 58:939.

Jackson, S. A., V. A. Rich, S. L. Ralston, and E. W. Anderson. 1985. Feeding behavior and feed efficiency of horses as affected by feeding frequency and physical form of hay. Proc. Eq. Nutr. Physiol. Symp. 9:78.

Jones, R. J., A. A. Seawright, and D. A. Little. 1970. Oxalate poisoning in animals grazing the tropical grass Setaria sphacelata. J. Aust. Inst. Agr. Sci. 36:41.

Kane, E., J. P. Baker, and L. S. Bull. 1979. Utilization of a corn oil supplemented diet by the pony. J. Anim. Sci. 40:1379.

Kelly, A. P., R. T. Jones, J. C. Gillick, and L. D. Sims. 1984. Outbreak of botulism in horses. Eq. Vet. J. 16:519.

Klendshoj, C., G. D. Potter, R. E. Lichtenwalner, and D. D. Householder. 1980. Nitrogen digestion in the small intestine of horses fed crimped or micronized sorghum or oats. Proc. Tex. Agr. Conf. College Station, Tex.: Texas A&M University.

Lawrence, L., K. J. Moore, H. F. Hintz, E. H. Jaster, and L. Wischover. 1987. Acceptability of alfalfa hay treated with an organic acid preparation. Can. J. Anim. Sci. 67:217.

Leonard, T. M., J. P. Baker, and J. Willard. 1975. Influence of distillers feeds on digestion in the equine. J. Anim. Sci. 40:1086.

Lewis, L. D. 1982. Feeding and Care of the Horse. Philadelphia: Lea & Febiger.

Ley, W. B. 1985. Mycotoxins in stored corn linked to fatal equine disease. Feedstuffs (January 28):7.

Morrison, F. B., J. G. Fuller, and G. Bohsted. 1919. Crushed versus whole oats for work horses. Wis. Agr. Exp. Stn. Bull. 302:63.

National Research Council. 1979. Interactions of Mycotoxins in Animal Production. Washington, D.C.: National Academy Press.

National Research Council. 1982. United States-Canadian Tables of Feed Composition. Washington, D.C.: National Academy Press.

Ott, E. A. 1972. Effect of Processing Feeds on Their Nutritional Value for Horses. Pp. 373–382 in Effect of Processing on the Nutritional Value of Feeds. Washington, D.C.: National Academy of Sciences.

Prichard, J. T., and J. L. Voss. 1967. Fetal ankylosis in horses associated with hybrid sudan pasture. J. Am. Vet. Med. Assoc. 150:871.

Prine, G. M. 1972. Perennial peanuts for forage. Proc. Soil Crop Soc. Fla. 32:33.

Reilly, P. J. 1981. The detection and identification of hordenine in the horse. Proc. 4th. Int. Conf. Control of the Use of Drugs in Racehorses. Melbourne, Australia.

Rich, V. A., J. P. Fontenot, and T. N. Meacham. 1981. Digestibility of animal, vegetable and blended fats by equine. Pp. 30–36 in Proc. 7th Eq. Nutr. Physiol. Soc. Symp. Warrenton, Va.

Ricketts, S. W., T. R. C. Greet, P. J. Glyn, C. D. R. Ginnett, E. P. McAllister, J. McCaig, P. H. Skinner, P. M. Webbon, D. L. Frape, G. R. Smith, and L. G. Murray. 1984. Thirteen cases of botulism in horses fed big bale silage. Eq. Vet. J. 16:515.

Riveland, N. R., D. O. Erickson, and E. W. French. 1977. An evaluation of oat varieties for forage. N. Dak. Farm Res. 35(1):19.

Rohweder, D. A., R. F. Barnes, and N. A. Jorgensen. 1978. Proposed hay grading standards based on laboratory analyses for evaluating quality. J. Anim. Sci. 47:747.

Schoeb, T. R., and R. J. Panciera. 1978. Blister beetle poisoning in horses. J. Am. Vet. Med. Assoc. 173:75.

Schurg, W. A. 1981. Alternative roughage utilization by horses. I. Evaluation of untreated or sodium hydroxide treated wheat straw in horse diets. Pp. 8–9 in Proc. 7th Eq. Nutr. Physiol. Soc. Symp. Warrenton, Va.

Schurg, W. A., D. L. Frei, P. R. Cheeke, and D. W. Holtan. 1977. Utilization of whole corn plant pellets by horses and rabbits. J. Anim. Sci. 45:1317.

Simms, B. T. 1951. Dallis grass poisoning. Auburn Vet. 8:22.

Smith, D., R. J. Bula, and R. P. Walgenbach. 1986. Forage Management, 5th. ed. Dubuque, Iowa: Kendall/Hunt Publ. Co.

Smith, G. R., and L. G. Murray. 1984. Laboratory confirmation of equine botulism. Vet. Rec. 114:75.

Sockett, D. C., J. C. Baker, and C. M. Stowe. 1982. Slaframine (Rhizoctonia leguminicola) intoxication in horses. J. Am. Vet. Med. Assoc. 181:606.

Sotola, J. 1937. The chemical composition and nutritive value of certain cereal hays as affected by plant maturity. J. Agr. Res. 54:399.

Taylor, T. H., and W. C. Templeton. 1976. Stockpiling Kentucky bluegrass and tall fescue forage for winter pasturage. Agron. J. 68:235.

Templeton, W. C. 1979. Forages for horses. Proc. Annu. Ky. Horsemen's Shortcourse 3:81.

Utley, P. R., W. C. McCormick, and R. S. Lowery. 1978. Weathered grass for wintering brood cows. Res. Rep. No. 293. Ga. Agr. Exp. Stn. Athens, Ga.: University of Georgia.

Van Kampen, K. R. 1970. Sudan grass and sorghum poisoning of horses: A possible lathrogenic disease. J. Am. Vet. Med. Assoc. 156:629.

Ward, G. M., L. H. Harbers, and J. J. Blaha. 1979. Calcium-containing crystals in alfalfa: Their fate in cattle. J. Dairy Sci. 62:715.

Webb, S. P., G. D. Potter, and K. J. Massey. 1985. Digestion of energy and protein by mature horses fed yeast culture. Proc. Eq. Nutr. Physiol. Symp. 9:64.

White, K. K., C. E. Short, H. F. Hintz, M. W. Ross, F. R. Lesser, P. F. Leids, J. E. Lowe, and H. F. Schryver. 1978. The value of dietary fat for working horses. II. Physical evaluation. J. Eq. Med. Surg. 2:525.

Wilcox, E. C. 1899. Ergotism in horses. Mont. Agr. Exp. Stn. Bull. No. 22:49

Willard, J. G., J. C. Willard, S. A. Wolfram, and J. P. Baker. 1977. Effect of diet on cecal pH and feeding behavior of horses. J. Anim. Sci. 45:87.

Woodroof, J. G. 1983. Peanuts: Production, Processing, Products, 3rd ed., J. G. Woodroof, ed., Westport, Conn.: AVI Publ.

GENERAL CONSIDERATIONS FOR FEEDING MANAGEMENT

Burke, P. R., G. D. Potter, W. C. McMullan, J. K. Krieder, T. R. Dotson, and D. S. Herring. 1981. Physiological effects of an anabolic steroid in the growing horse. Paper presented at Symp. Anabolic Steroids Eq. Med. Columbia, Mo.: University of Missouri.

Matsuaka, T. 1976. Evaluation of monensin toxicity in the horse. J. Am. Vet. Med. Assoc. 169:1098.

O'Conner, J. J., M. C. Stillions, W. A. Reynolds, W. H. Lindenheimer, and S. C. Matlesden. 1973. Evaluation of boldenone undecylenate as an anabolic agent in horses. Can. Vet. J. 14:154.

Squires, E. L., G. E. Todter, W. E. Berndtson, and B. W. Pickett. 1982. Effect of anabolic steroids on reproductive function of young stallions. J. Anim. Sci. 54:576.

Squires, E. L., J. L. Voss, J. M. Maher, and R. K. Shideler. 1985. Fertility of young mares after long term anabolic steroid treatment. J. Am. Vet. Med. Assoc. 186:58.

FEED COMPOSITION

Fonnesbeck, P. V. 1981. Estimating digestible energy and TDN for horses with chemical analyses of feeds. J. Anim. Sci. 53(Suppl. 1):241 (Abstract).

RESEARCH FINDINGS ON COMPOSITION OF MILK (Appendix Table 1)

Anwer, M. S., R. Cronwall, W. E. Chapman, and R. D. Klenz. 1975. Glucose utilization and contribution to milk components in lactating ponies. J. Anim. Sci. 41:568.

Baucus, K. L., S. L. Ralston, G. Rich, and E. L. Squires. 1987. The effect of dietary copper and zinc supplementation on composition of mares milk. Pp. 179–184 in Proc. 10th Eq. Nutr. Physiol. Soc. Symp. Fort Collins, Colo.: Colorado State University.

Bouwman, H., and W. van der Schee. 1978. Composition and production of milk from Dutch warmblooded saddle horse mares. Z. Tierphysiol. Tierernaehr. Futtermittelk. 40:39.

Breedveld, L., S. G. Jackson, and J. P. Baker. 1987. The determination of a relationship between copper, zinc, and selenium levels in mares and those in the foals. Pp. 159–164 in Proc. 10th Eq. Nutr. Physiol. Soc. Symp. Fort Collins, Colo.: Colorado State University.

Doreau, M., S. Boulot, W. Martin-Rosset, and H. DuBroeucq. 1986. Milking lactating mares using oxytocin: Milk volume and composition. Reprod. Nutr. Dev. 26-1.

Duisembaev, K. I., and B. R. Akimbekov. 1982. Variation of milk yield and its relationship with milk composition of mares at a koumiss farm. Sbornik Nauchnvkh Trudov. Kazakhskii Nauchno-Issledovatel'skii Teknological Institut Ovtsevodstva (as cited in J. Dairy Sci. 46:1984) (Abstract).

Fedotov, P., and B. Akimbekov. 1983. Increasing milk production of Kushum mares. Konevodstvo i Konnyi Sport. (11):6–7 (as cited in J. Dairy Sci. 46:264) (Abstract).

Gibbs, P. D., G. D. Potter, R. W. Blake, and W. C. McMullan. 1982. Milk production of quarter horse mares during 150 days of lactation. J. Anim. Sci. 54:496.

Kulisa, M. 1986. Some components of mare milk. Paper presented at 37th Annu. Meet. Eur. Assoc. Anim. Prod., Budapest, Hungary, September 1–4, 1986. Summaries, vol. 2. Commission on horse production (as cited in J. Dairy Sci. 49:806) (Abstract).

Lukas, V. K., W. W. Albert, F. N. Owens, and A. Peters. 1972. Lactation of Shetland mares. J. Anim. Sci. 34:350 (Abstract).

Meadows, D. G., G. D. Potter, W. B. Thomas, J. Hesby, and J. G. Anderson. 1979. Foal growth, milk production and milk composition from mares fed combinations of soybean meal or urea supplements. J. Anim. Sci. 49(Suppl. 1):247.

Neseni, R., E. Flade, G. Heidler, and H. Steger. 1958. Milchleistung und Milchzusammensetzung von Stuten im Verlaufe der Laktation. Arch. Tierzucht. 1:91.

Neuhaus, U. 1959. Milch und Milchgewinnung von Pferdestuten. Z. Tierzucht. 73:370.

Oftedal, O. T., H. F. Hintz, and H. F. Schryver. 1983. Lactation in the horse: Milk composition and intake by foals. J. Nutr. 113:2169.

Pagan, J. D., and H. F. Hintz. 1986. Composition of milk from pony mares fed various levels of digestible energy. Cornell Vet. 76:139.

Schryver, H. F., O. T. Ofteda, J. Williams, N. F. Cymbaluk, D. Antczak, and H. F. Hintz. 1986a. A comparison of the mineral composition of milk of domestic and captive wild equids (E. przewalski, E. zebra, E. burchelli, E. caballus, E. asinus). Comp. Biochem. Physiol. 85A:233.

Schryver, H. F., O. T. Oftedal, J. Williams, L. V. Soderholm, and H. F. Hintz. 1986b. Lactation in the horse: The mineral composition of mare milk. J. Nutr. 116:2142.

Smoczynski, S., and R. Tomczynski. 1982. Composition of mare's milk. I. The first 10 days of lactation. Badania skladu chemicznego mleka klaczy. I. Pierwsze dziesiec dni laktacji. Zeszyty Naukowe Adademii Rolniczo-Technicznej w Olsztynie, Technologia Zywnosci. (17):77–83 (as cited in Dairy Sci. 45:777) (Abstract).

Ullrey, D. E., R. D. Struthers, D. G. Hendricks, and B. E. Brent. 1966. Composition of mare's milk. J. Anim. Sci. 25:217.

Ullrey, D. E., E. T. Ely, and R. L. Covert. 1974. Iron, zinc, and copper in mare's milk. J. Anim. Sci. 38:1276.

Appendix Table

APPENDIX TABLE 1 Research Findings on Composition of Milk

Time After Foaling	Number of Animals	Total Solids (%)	Energy (kcal/100 g)	Protein (%)	Fat (%)	Lactose (%)	Concentration (µg/g of fluid milk)							References
							Calcium	Phosphorus	Magnesium	Potassium	Sodium	Copper	Zinc	
1–4 weeks	10	11.6	59	3.1	2.1	5.9	1,212	433	88	773	246	0.46	3.2	Ullrey et al. (1966, 1974)
5–8 weeks	10	11.1	55	2.5	1.9	5.9	1,008	305	56	505	196	0.24	2.5	
9–17 weeks	10	10.2	50	2.0	1.3	6.5	661	230	41	388	168	0.22	2.2	
1–4 weeks	5	11.1	56	2.3	1.6	6.8	1,223	811	101	586	194	0.62	2.6	Schryver et al. (1986a,b)
5–8 weeks	5	10.5	50	1.9	1.3	6.9	894	596	66	420	167	0.38	1.9	
9–17 weeks	2						786	557	49	370	137	0.21	1.8	Oftedal et al. (1983)
1–4 weeks	14	10.9		2.5	1.5									Gibbs et al. (1982)
5–8 weeks	14	10.5		2.1	1.4									
9–21 weeks	14	10.2		1.9	1.0									
2–12 weeks	22[a]	10.4	45	2.2	0.8	6.6	1,220	660						Pagan and Hintz (1986)
1–10 weeks	5[a]						857	418	77	380	127	0.37	1.7	Schryver et al. (1986a)
1–4 weeks	20											0.17	1.8	Breedveld et al. (1987)
5–8 weeks	20											0.17	1.8	
1–4 weeks	20	9.8					915	502	85			0.32	2.2	Baucus et al. (1987)
1 week	6[a]	9.8		3.1	1.6									Lukas et al. (1972)
6 weeks	6[a]	9.6		2.1	1.4									
18 weeks	6[a]	9.8		1.9	1.8									
1–4 weeks	5	11.6		2.5	2.0		1,240	780						Bouwman and van der Schee (1978)
1–12 weeks	28	10.4		2.0	1.6									Meadows et al. (1979)
6–8 weeks	24	11.2	47	2.6	1.9	6.2	1,179	926		887	203			Neseni et al. (1958)
6–8 weeks	3[a]			2.1	1.4	6.9								Anwer et al. (1975)
	10			3.3	1.6	6.2								Doreau et al. (1986)
Day 10	Review[b]	10.0		2.1	1.3	6.2	1,050	715		685	170			Neuhaus (1959)
	26	11.0		3.1	1.9	6.4								Smoczynski and Tomczynski (1982)
Average of 6-month lactation	110			2.0	1.4	6.1		394	29			0.25	0.9	Kulisa (1986)
Average of 6-month lactation		10.2		1.9	1.6	6.6								Fedotov and Akimbekov (1983)
Average of 6-month lactation	25			2.0	1.6	6.4								Duisembaev and Akimbekov (1982)
Summary														
1–4 weeks		10.7	58	2.7	1.8	6.2	1,200	725	90	700	225	0.45	2.5	
5–8 weeks		10.5	53	2.2	1.7	6.4	1,000	600	60	500	190	0.26	2.0	
9–21 weeks		10.0	50	1.8	1.4	6.5	800	500	45	400	150	0.20	1.8	

[a] Ponies.
[b] Review of many studies; values are averages obtained from Neuhaus (1959).

Index

Other Titles in the Series

Nutrient Requirements of Dairy Cattle, 6th Rev. Ed., Update 1989	0-309-03826-X
Nutrient Requirements of Swine, 9th Rev. Ed., 1988	0-309-03779-4
Nutrient Requirements of Cats, Rev. Ed., 1986	0-309-03682-8
Nutrient Requirements of Sheep, 6th Rev. Ed., 1985	0-309-03596-1
Nutrient Requirements of Dogs, Rev. Ed., 1985	0-309-03496-5
Nutrient Requirements of Poultry, 8th Rev. Ed., 1984	0-309-03486-8
Nutrient Requirements of Beef Cattle, 6th Rev. Ed., 1984	0-309-03447-7
Nutrient Requirements of Warmwater Fishes and Shellfishes, Rev. Ed., 1983	0-309-03428-0
Nutrient Requirements of Mink and Foxes, Rev. Ed., 1982	0-309-03325-X
Nutrient Requirements of Coldwater Fishes, 1981	0-309-03187-7
Nutrient Requirements of Goats, 1981	0-309-03185-0
Nutrient Requirements of Laboratory Animals, 3rd Rev. Ed., 1978	0-309-02767-5
Nutrient Requirements of Nonhuman Primates, 1978	0-309-02786-1
Nutrient Requirements of Horses, 4th Rev. Ed., 1978	0-309-02760-8
Nutrient Requirements of Rabbits, 2nd Rev. Ed., 1977	0-309-02607-5

Related Publications

Predicting Feed Intake of Food-Producing Animals, 1987	0-309-03695-X
Vitamin Tolerance of Animals, 1987	0-309-03728-X

Further information, additional titles of
Board on Agriculture publications, and prices are
available from the National Academy Press,
2101 Constitution Ave., NW, Washington, DC 20418.